T0238458

BestMasters

Springer awards „BestMasters" to the best master's theses which have been completed at renowned universities in Germany, Austria, and Switzerland.

The studies received highest marks and were recommended for publication by supervisors. They address current issues from various fields of research in natural sciences, psychology, technology, and economics.

The series addresses practitioners as well as scientists and, in particular, offers guidance for early stage researchers.

Bettina Basel

Dipolar Correlation Spectroscopy

Higher-Order Correlation Terms in Three-Spin Double Electron-Electron Resonance (DEER)

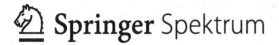

Springer Spektrum

Bettina Basel
Erlangen, Germany

BestMasters
ISBN 978-3-658-09190-3   ISBN 978-3-658-09191-0 (eBook)
DOI 10.1007/978-3-658-09191-0

Library of Congress Control Number: 2015932771

Springer Spektrum
© Springer Fachmedien Wiesbaden 2015
This work is subject to copyright. All rights are reserved by the Publisher, whether the whole or part of the material is concerned, specifically the rights of translation, reprinting, reuse of illustrations, recitation, broadcasting, reproduction on microfilms or in any other physical way, and transmission or information storage and retrieval, electronic adaptation, computer software, or by similar or dissimilar methodology now known or hereafter developed.
The use of general descriptive names, registered names, trademarks, service marks, etc. in this publication does not imply, even in the absence of a specific statement, that such names are exempt from the relevant protective laws and regulations and therefore free for general use.
The publisher, the authors and the editors are safe to assume that the advice and information in this book are believed to be true and accurate at the date of publication. Neither the publisher nor the authors or the editors give a warranty, express or implied, with respect to the material contained herein or for any errors or omissions that may have been made.

Printed on acid-free paper

Springer Spektrum is a brand of Springer Fachmedien Wiesbaden
Springer Fachmedien Wiesbaden is part of Springer Science+Business Media
(www.springer.com)

This book is dedicated to my chemistry teacher Markus Eiber without whom I would not have ended up in the field of chemistry after all.

Acknowledgment

First of all, I would like to thank *Prof. Gunnar Jeschke* for giving me the chance to do my master thesis in his group and being a great boss. He always took the time for discussions and gave friendly and considered advice on how I should proceed.

Moreover, I am indebted to *Andrin Doll* for his continuous help, for all the explanations and extra lessons he gave me and the patience he had with me. Furthermore, I am much obliged to *Prof. Dirk Guldi* for establishing the contact to ETH, supporting me in many aspects during all my studies and his pleasant advice.

This thesis would have not been possible without the model compounds. Therefore I thank *Patrick Wilhelm, Ludmila Ziegler* and *Prof. Helma Wennemers* for synthesizing the poly(proline), *Luca Garbuio* for preparing the poly(proline) samples and *Muhammad Sajid, Miriam Schulte* and *Prof. Adelheid Godt* for synthesizing the nitroxide radicals.

A big thank goes to *Dr. Maxim Yulikov* and *Dr. Vidmantas Kalendra* for all the fun and the good time we spent together.

I would also like to thank *Kristina Comiotto Schiavetta* and *Tona von Hagens* for interesting conversations and their cordiality towards me.

Besides, I thank all present and former members, visitors and students of the EPR group at ETH for receiving me warmly and the enjoyable time.

Finally, I want to express my gratitude to the *Erasmus programme*, the *Studienstiftung des deutschen Volkes* and *my parents* for funding my stay at ETH. Besides the financial supports *my parents* have been a constant source of love and aid through all my studies. I greatly appreciate this.

Bettina Basel

Abstract

This master thesis presents experimental conditions which allow to extract the three-spin contribution precisely from experimental DEER signals. Broadband chirp pulses, generated by a fast arbitrary waveform generator (AWG), were used as DEER pump pulses in combination with a homebuilt, highly overcoupled microwave resonator. This experimental setup was tested on triply labeled nitroxide radicals and a Gd(III)-nitroxide radical. To evaluate the quality of the extracted two-spin and three-spin contribution, simulations were performed.

The nitroxide triradicals showed orientation selection. This was examined by a newly developed double chirp pulse consisting out of two separate chirps. This procedures provided the possibility to generate comparable DEER data at observation positions uniformly distributed all over the nitroxide spectrum. From this data the conclusion was drawn that three spin labels of the same type are not suitable for generating DEER data from which the three-spin contribution can be extracted with the assumptions that were published by *Jeschke, G. et al.* in 2009 [1]. In all experiments, the orientation selection was found to be more significant in the three-spin than in the two-spin contribution.

The observation selection problem could be avoided by changing the model compound to a Gd(III)-nitroxide radical. The observation frequency was set at the Gd(III) and the whole nitroxide spectrum was pumped with the chirp pulse. Due to echo reduction the observation frequency had to be moved away from the Gd(III) central transition. With this setup, orientation selection due to both observation frequency and pumping bandwidth could be avoided and the experimental and simulated two-spin and three-spin contributions matched.

In almost all experiments with chirp pulses, the phase-corrected DEER data showed significant contributions in its imaginary part. The origin of this unusual observation was examined by test experiments on a nitroxide diracidal. It could be excluded that the information contained in the real part of the phase corrected DEER signal is affected by the non-zero imaginary part.

Contents

Symbols and Abbreviations

Symbols

$\mathbf{A}$	Hyperfine tensor
a_{iso}	Isotropic hyperfine coupling constant
a_{kl}	Amplitude of the transition between the k and l state
$\mathbf{B_0}$	Static magnetic field vector
$\mathbf{B_1}$	Magnetic field vector of the applied m.w. radiation
B_1	M.w. amplitude
$\mathbf{B_{eff}}$	Effective magnetic field vector
β	Flip angle of an applied m.w. pulse
β_{e}	Bohr magneton ($9.27400949(80)*10^{-24}$ J/T)
β_{n}	Nuclear magneton ($5.05078324(13)*10^{-27}$ J/T
$\mathbf{B_{local}}$	Local magnetic field vector
$B(t)$	Background factor of the DEER signal
C	Capacitance
$\mathbf{D}$	Zero-field interaction tensor
d	Dimensionality of the homogeneous distribution of nanoobjects on which DEER is performed
δ	EPR analog of the 'chemical shift'
Δ	Modulation depth
$\Delta\nu$	Bandwidth of a resonator

$\Delta\phi_{ee}$	Out of phase angle
e	Elementary charge $(1.602176565(35)*10^{-19}$ C$)$
ε	Coupling coefficient
η	Filling factor of the resonator
$F(\omega/2\pi)$	Dipolar spectra = Fourier transformation of a form factor
$F(t)$	Experimental form factor = phase and background corrected DEER signal
$F_3(t)$	Form factor of a three-spin system
$f_{kl}(t,\theta_{kl},r_{kl})$	Dipolar evolution function of the spin pair (k,l)
$\mathbf{g}$	Tensor form of the effective electron Zeeman factor
g	Effective electron Zeeman factor for an isotropic system
g_e	Free-electron Zeeman factor $(2.0023193043617(15))$
g_{eff}	Effective electron Zeeman factor for an anisotropic system
g_n	Nuclear Zeeman factor
h	Planck constant $(4.135667516(91)*10^{-15}$ eV s $= 6.62606957(29)*10^{-34}$ J s$)$
$\mathcal{H}_0$	Static spin Hamiltonian
$\mathcal{H}_1$	Time-independent oscillatory Hamiltonian
$\hbar$	Reduced Planck constant $(6.58211928(16)*10^{-15}$ eV s $= 1.054571726(47)*10^{-34}$ J s$)$
$\mathcal{H}_{DD}$	Hamiltonian of the electron-nuclear dipole-dipole coupling
$\mathcal{H}_{dd}$	Hamiltonian that describes the dipole-dipole coupling
$\mathcal{H}_{exch}$	Hamiltonian that describes the exchange coupling
$\mathcal{H}_{EZ}$	Electron Zeeman Hamiltonian
$\mathcal{H}_F$	Hamiltonian of the Fermi contact interaction
$\mathcal{H}_{HF}$	Hamiltonian that describes the hyperfine couplings

$\mathcal{H}_{NN}$	Hamiltonian that describes spin-spin interactions
$\mathcal{H}_{NQ}$	Nuclear quadrupole Hamiltonian
$\mathcal{H}_{NZ}$	Nuclear Zeeman Hamiltonian
$\mathcal{H}_{ZFS}$	Hamiltonian that describes the zero-field splitting
$\mathbf{I}$	Nuclear spin vector operator
I	Complex current amplitude
$I_t(t)$	Alternate current
$\mathbf{J}$	Exchange coupling tensor
k_{dec}	Rate constant of the decay of the intermolecular couplings that contribute to the Background factor of the DEER signal
$K_{\nu\varepsilon}$	Voltage-reflection coefficient
$K_{P\varepsilon}$	Power-reflection coefficient
L	Inductance
λ	Modulation depth parameter, inversion efficency of the pumped spins
μ	Electron magnetic moment vector
$\mathbf{M}$	Macroscopic magnetization vector
m_e	Electron rest mass ($9.11*10^{-31}$ kg)
m_s	Secondary spin quantum number or magnetic quantum number
μ_0	Vacuum permeability, $4\pi \times 10^{-7}$ N A^{-2}
n	Coupling between a transmission line and a resonator
N_v	The number of dipoles per unit volume
$\nu_{resonator}$	Resonance frequency of the resonator
ω_0	Larmor frequency

ω_1	Precession frequency about the applied m.w. field vector
ω_{dd}	Dipole-dipole coupling
ω_{ee}	Electron-electron coupling
ω_{eff}	Nutation frequency about the effective field
ω_{mw}	Frequency of the applied microwave radiation
Ω_S	Resonance offset
ω_{spin1}	Resonance frequency of the observer spin ($=$ observation frequency)
ω_{spin2}	Resonance frequency of pumped spin ($=$ pump frequency)
$P(\omega/2\pi)$	Dipolar spectra of the two-spin contribution $=$ Fourier transformation of the time domain signal of the two-spin contribution
$P(t)$	Time-domain signal of the two-spin contribution
P	General two-spin contribution (not specified which domain)
P_{mw}	M.w. power
Q	Adiabaticity
Q_L	Loaded quality factor
R	Resistance
$\mathbf{S}$	Spin angular momentum operator
S	Primary spin quantum number
σ	Density operator
σ_{kl}	Matrix element of the density operator
$T(\omega/2\pi)$	Dipolar spectra of the three-spin contribution $=$ Fourier transformation of the time domain signal of the three-spin contribution
$T(t)$	Time-domain signal of the three-spin contribution

T	Dipolar coupling tensor
T	General three-spin contribution (not specified which domain)
T	Delay time in a pulse EPR experiment
τ	Evolution time in a pulse EPR experiment
θ_{dd}	Angle between the inter-spin vector and the external magnetic field
t_p	Duration of an applied m.w. pulse
$V(t)$	Phase corrected experimental DEER signal
V	Complex voltage amplitude
$V_3(t)$	DEER signal of system of a three-spin system
V_c	Effective volume of the resonator
V_E	Echo voltage induced in the resonator of inductance L
V_{E0}	Echo signal at the detector for the critically coupled case
$V_{E\varepsilon}$	Echo signal at the detector for a coupling coefficient ε
$V_{pair}(t)$	DEER signal of a two-spin system
$V_t(t)$	Alternate voltage
w_{kl}	Probability of the transition between the k and l state
Z	Impedance

Abbreviations

AC	Alternate current
AWG	Arbitrary waveform generator
DEER	Double electron-electron resonance

EPR	Electron paramagnetic resonance
ESE	Echo-detected field sweep
FM	Frequency-modulated
FRET	Förster resonance energy transfer
IF	Intermediate frequency
LO	Local oscillator
m.w.	Microwave
MPFU	Microwave pulse forming unit
NMR	Nuclear magnetic resonance
norm.	Normalized
obs	Observation
PELDOR	Pulsed electron-electron double resonance
I	In-phase component
Q	Quadrature component
UWB	Ultra-wide band

List of Tables

List of Figures

1 A General Introduction

1.1 Basic Principles of Electron Paramagnetic Resonance

Molecular absorption spectra are obtained by measuring the attenuation versus frequency of a beam of electromagnetic radiation as it passes through a sample of matter. In such kinds of spectroscopy, the electric field component of the radiation interacts with the electric dipole moment of the molecule. In magnetic resonance spectroscopy, however, the interaction of magnetic dipole moments with the oscillating magnetic component of electromagnetic radiation is measured. Molecular magnetic moments may either arise from the nuclei (nuclear magnetic resonance, NMR) or the electrons (electron paramagnetic resonance, EPR) of a system. [2]

Each electron possesses an intrinsic magnetic dipole moment. However, if electrons occur in pairs, like in most systems, the net moment is zero. Therefore only species containing one or more unpaired electrons can be studied by EPR spectroscopy, such as *compounds with transition metal ions, free radical intermediates, paramagnetic impurities in crystals and glasses* and *stable free radicals like nitroxides*. In almost all cases encountered in EPR spectroscopy, the electron magnetic dipole is caused by the spin angular momentum with only a small contribution from orbital motion. [2, 3]

1.1.1 Magnetic Moments and Angular Momenta

An electron possesses an intrinsic angular momentum $\hbar\mathbf{S}$ which gives rise to a magnetic moment

$$\mu = -g\beta_{\mathrm{e}}\mathbf{S}. \tag{1.1}$$

With the Bohr magneton defined by

$$\beta_{\mathrm{e}} = \frac{e\hbar}{2m_{\mathrm{e}}}. \tag{1.2}$$

g_e is the free-electron Zeeman factor (2.0023193043617(15)). It is required to account for deviations of the behavior of a quantum object from the behavior of a classical charged particle. e is the elementary charge (1.602176565(35)* 10^{-19} C), m_e is the electron rest mass (9.11*10^{-31} kg) and **S** is the spin angular momentum operator. [2, 3]

In quantum mechanics the allowed values of the magnitude of any angular momentum originating from its operator **J** are given by $[J(J+1)]$ where J is the primary angular-momentum quantum number ($J = 0, \frac{1}{2}, 1, \frac{3}{2}, ...$). All angular momenta and their components are given in units of $\hbar$. Thus for a single electron **S** has the quantum number $S = \frac{1}{2}$, for systems of two or more unpaired electrons S is $1, \frac{3}{2}, 2, ...$ For paramagnetic ions, especially those of the transition ions, states with $S > \frac{1}{2}$ are common. [2]

1.1.2 Electron Zeeman Effect

The energy differences studied in EPR spectroscopy are mainly caused by the interaction of the magnetic moment of the unpaired electrons in the sample and an external magnetic field, $\mathbf{B_0}$. This interaction is called the Zeeman effect. If an electron is placed in a magnetic field, it has a state of lower energy when the moment of the electron, μ, is aligned with the magnetic field and a higher energy state when μ is aligned against the magnetic field. The two states are designated by the projection of the electron spin, m_s, (called the secondary spin quantum number or magnetic quantum number) on the direction of the magnetic field. For a $S = \frac{1}{2}$ system, like a single electron, the parallel state is indicated with $m_s = -\frac{1}{2}$ and the antiparallel with $m_s = +\frac{1}{2}$. For systems with higher S values ($1, \frac{3}{2}, 2, ...$) placed in a magnetic field there are more than two separated states indicated by m_s. Generally m_s can take the values $(-S), (-S+1), ..., 0, ..., (S-1), S$. This leads to a set of energies, $U(m_s)$, which are sometimes referred to as *electronic Zeeman energies*. [2, 4]

$$U(m_s) = g_e \beta_e B m_s \tag{1.3}$$

For a transition between the Zeeman levels the conservation of angular momentum imposes a selection rule of $|\Delta m_s| = 1$. [2]

1.1.3 Characteristics of the Spin Systems

1.1.3.1 The g Factor

As already indicated g_e is only valid for free electrons. If an electron is part of a system g_e must be replaced by variable g factor which is called effective electron Zeeman factor. This is due to local fields that add vectorial to the applied external field. Thus the effective field, $\mathbf{B_{eff}}$, that is felt by an electron varies from the external magnetic field, $\mathbf{B_0}$. [2]

$$\mathbf{B_{eff}} = \mathbf{B_0} + \mathbf{B_{local}} \tag{1.4}$$

There are two types of local fields, those that are induced by $\mathbf{B_0}$ and those that are permanent and independent of $\mathbf{B_0}$. As those local fields are hard to measure, it is much easier to just keep the external magnetic field, $\mathbf{B_0}$, for calculations and replace g_e by a variable g factor instead. If one considers only induced local fields that depend on $\mathbf{B_0}$, one can write equation 1.4 as

$$\mathbf{B_{eff}} = (1 - \delta)\mathbf{B_0} = \frac{g}{g_e}\mathbf{B_0}. \tag{1.5}$$

δ is the EPR analogon of the 'chemical shift' parameter δ_n used in NMR spectroscopy. However this expression is only valid for systems that behave isotropically. With anisotropic systems, variability of g with orientation relative to $\mathbf{B_0}$ is required: [2]

$$g_x \neq g_y \neq g_z \tag{1.6}$$

1.1.3.2 Characteristics of Dipolar Interactions

As mentioned before, besides those local fields that are induced by $\mathbf{B_0}$ there exists also another form of local fields that is independent from $\mathbf{B_0}$. These permanent locals fields are caused by neighbouring dipoles. The magnitude and direction of the local-field contribution depend on the spin state of the center containing the dipole. If the observed electron is interacting with a neighbouring nuclear dipole moment the interaction is called *hyperfine interaction*. If it is interacting with another unpaired electron this is called *electron-electron dipolar interaction*. As the electron magnetic moment is much larger than that of nuclei the latter usually dominates the spectral features. [2]

1.2 Structure Determination of Bio-Macromolecules

It is essential to know the structure of bio-macromolecules like enzymes, receptors or nucleic acids to understand the way they function and to design pharmaceuticals. Currently NMR spectroscopy, electron microscopy and X-ray crystallography are the standard methods for determining the structure of bio-macromolecules at atomic resolution. However these methods cannot be applied on every bio-macromolecule. For example, not all proteins can be crystallized and others might not crystallize in their biologically relevant conformation. Proteins that occur in biological membranes are especially hard to crystallize. However membrane proteins can form 2-dimensional arrays in the membrane that can be investigated by electron microscopy. The drawback of this technique is that the resolution is in most cases limited to 10-20 Å. The advantage of NMR is that the molecules can be investigated in solution and crystallization is not needed. On the other side the structure determination by NMR is based on the analysis of numerous nuclear spin-spin interactions. This is only efficient for small or medium-sized bio-macromolecules.

To overcome some of the experimental limitations of the three standard methods, probe-based techniques like FRET (Förster resonance energy transfer) and DEER (double electron-electron resonance) provide complementary information. Both do not require crystallization, provide specific distances in the Angström to nanometer range, are limited neither by size nor by weight and can be applied on flexible and less ordered regions. FRET allows distance measurements in the range of 25-60 Å, extending up to 100 Å in favorable cases. To perform a FRET measurement the bio-macromolecule has to be labeled with a donor and an acceptor fluorophore that are chosen such that the emission spectrum of the donor and the excitation spectrum of the acceptor overlap. In contrast to this, DEER, which is also known as PELDOR (pulsed electron-electron double resonance), measures the interaction of paramagnetic centers: the distance dependent electron-electron dipolar interaction. Normally the investigated bio-macromolecules have to be mutated with an EPR active group, called spin label, for this.

FRET can be performed on freely diffusing molecules in solution, while DEER is mostly performed in glassy frozen solutions. For FRET the interacting labels need to be dissimilar, as donor and acceptor are needed, whereas they may be identical for DEER. The most widely used approach for labeling a molecule for DEER is to incorporate a cysteine residue at the desired position and subsequently conjugate it to a nitroxide spin label. Alternatively chelated lanthanoide complexes can be introduced as spin labels. The advantage of DEER is that the diagmagnetic core of the investigated system is EPR silent and does not disturb the signal. Therefore distances in the range 1.5 to 8 nm in bio-macromolecules of arbitrary size can be studied by DEER. [5–8]

1.3 Goal of the Thesis

In this thesis I will focus on higher-order correlation terms in the DEER signal. Those appear if more than two coupled spins are present in the investigated system. They give rise to signal contributions of combination frequencies. If the data analysis procedure used for spin pairs is applied to those signals, additional distances that do not correspond to the real inter-spin distances will be found. Those are referred to as *ghost contributions.* [9] In 2013 *von Hagens, T. et al.* presented the *power scaling* approach in order to suppress ghost distances in the analysis of the DEER signal. [10] Besides the danger of misinterpretations of the DEER signal, the higher-order correlations are also a source of information. In 2009 *Jeschke, G. et al.* proved that it is, in principle, possible to extract not only inter-spin distances, but also inter-spin angles by fitting a geometry model to an experimental DEER signal that was measured on a three-spin system. [1] Fitting a geometry model is of course only possible for molecules with not too complex structures or if an idea about how the structure might look like has already been developed by pre-examinations with other techniques. Thus, this approach is neither satisfying and nor applicable for structure determination of unknown bio-macromolecules. The optimum solution would be the development of a new analysis procedure to extract the angular information out of the higher-order terms. However, those scale with the power of $(n\text{-}1)$ of the spin inversion efficiency, whereat n is the number of frequencies that are combined in the particular term of higher-order. Thus, the inversion efficiency needs to be very high to extract this terms with sufficient precision.

The goal of this thesis is to find experimental conditions under which it is possible to extract the second-order term, named *three-spin contribution,* from the DEER signal precisely. To achieve high inversion efficiencies over a broad spectral range *Doll, A. et al.* in 2013 interfaced a fast arbitrary waveform generator (AWG) to a commercial EPR spectrometer. [11] This setup was used in combination with a homebuilt broadband resonator for performing DEER experiments with sufficient inversion efficiencies. It was tested on triply labeled nitroxide radicals and a Gd(III)-nitroxide radical to find the optimum labels and procedure of performing the measurements.

1.4 Structure of the Thesis

This work is structured as follows. The basic theory of EPR is introduced in the *Introduction*, chapter 1. The *Theory* part, chapter 2, focuses on the relevant theoretical knowledge about EPR. Therefore, it gives insights into its classical and quantum mechanical description. Moreover, it provides fundamentals about the used methods, complications that might appear and the expected signals. Besides this the investigated molecules and the used instrumentations are discussed in detail. In the *Scope*, chapter 3, a detailed preface of all performed experiments is given based on the theory chapter. As the goal of my work was finding optimum experimental conditions, the *Materials and Methods* section, chapter 4, is an important part and documents, besides the used chemicals and materials, the main considerations and reasons for choosing the tested conditions. Furthermore, the procedures of data analysis used for this work are introduced. In chapter 5, *Results and Discussion*, the outcome of the performed experiments are described. The main focus is put on evaluating the generated measurement data with respect to quality and reliability of the extraction of the three-spin contribution. Therefore, simulations of the expected three-spin contribution are performed to compare them to the experimental data. Moreover, test measurements on a reference compound are presented to appraise unknown artifacts.

2 Theory

2.1 A Classical Description of the EPR Experiment

2.1.1 The Bulk Magnetization

To describe the EPR experiment in a classical way, the interaction of the macroscopic bulk magnetization with the applied microwave frequency has to be regarded. If one considers a large ensemble of non-interacting classical magnetic moments in a high magnetic field ($\sim$ 1 T) at low temperatures ($\sim$ 1 K), the resulting macroscopic magnetization $\mathbf{M}$ would be approximately equal to $N_v\mu$. N_v is the number of dipoles per unit volume. [2, 12]

If the classical equation of motion is applied for a magnetic moment with an angular momentum in a magnetic field to the bulk magnetization, one gets

$$\frac{d\mathbf{M}}{dt} = \mathbf{M} \times \frac{-g\beta_e}{\hbar}\mathbf{B}(t). \tag{2.1}$$

Equation 2.1 describes a precession of the magnetization vector about the axis of the static external magnetic field vector (in the following set as the z-axis of the coordinate system). The precessional frequency is called the *Lamor frequency*, ω_0. [12]

$$\omega_0 = \frac{g_e\beta_e\mathbf{B}_0}{\hbar} \tag{2.2}$$

2.1.2 The Rotating Frame

At equilibrium the magnetization vector is aligned parallel to the axis of the static external magnetic field vector. To move it away from its equilibrium position one can apply an alternating magnetic field, $\mathbf{B}_1$, oscillating with the Larmor frequency of the observed spins perpendicular to the external magnetic field vector. Usually linearly polarized fields are used for this purpose. For convenience, the direction of $\mathbf{B}_1$ is defined as the x-axis of the laboratory coordinate system. In this case the linear polarized alternating magnetic field can be described as a superposition of two circularly polarized alternating magnetic fields: A right-handed polarized fields,

$$B_{1x}^{(r)}(t) = B_1 \cos(\omega_{mw}t), \quad B_{1y}^{(r)}(t) = B_1 \sin(\omega_{mw}t), \quad B_{1z}^{(r)}(t) = 0, \quad (2.3)$$

and a left-handed polarized field,

$$B_{1x}^{(l)}(t) = B_1 \cos(\omega_{mw}t), \quad B_{1y}^{(l)}(t) = -B_1 \sin(\omega_{mw}t), \quad B_{1z}^{(l)}(t) = 0, \quad (2.4)$$

whereat ω_{mw} is the frequency of the microwave radiation. The right-handed polarized fields follows the precession of the magnetization vector, whereas the left-handed polarized field precesses in the opposite direction. Therefore, only the right-handed polarized field influences the magnetization vector distinctly. [12]

However, with a time-dependent field like this, equation 2.1 cannot be solved analytically. The solution to this problem is to define our coordinate system such that it rotates with ω_{mw} anti-clockwise about the z-axes. Then the time dependency of $B_1^{(r)}$ is removed. Such a coordinate system is called a *rotating frame*. [12]

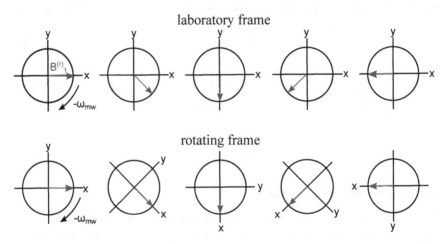

Figure 2.1: Sketch of the movement of a right-handed polarized magnetic field, $B_1^{(r)}$, in a laboratory and a rotating frame. [13]

In the rotating frame the magnetization vector precesses with the difference between its Larmor frequency, ω_0, and the frequency of the applied microwave radiation, ω_{mw}. The frequency difference is called the *resonance offset*, Ω_S. [12]

$$\Omega_S = \omega_0 - \omega_{\mathrm{mw}} \tag{2.5}$$

2.1.3 The Bloch Equations

In a rotating frame equation 2.1 can be written as a set of linear equations:

$$\frac{\mathrm{d}M_x}{\mathrm{d}t} = -\Omega_S M_y,$$

$$\frac{\mathrm{d}M_y}{\mathrm{d}t} = \Omega_S M_x - \omega_1 M_z,$$

$$\frac{\mathrm{d}M_z}{\mathrm{d}t} = -\omega_1 M_y \tag{2.6}$$

$$\text{with } \omega_1 = \frac{g_x \beta_e B_1}{\hbar} \tag{2.7}$$

However, those equations cannot be a complete description of the motion of the magnetization vector in a magnetic field, as they do not predict that after some time the magnetization vector will reach its equilibrium position and align with the external magnetic field. To explain this observation relaxation processes have to be taken into account. The z-component of the magnetization vector relaxes with the so-called longitudinal relaxation time, T_1, while the x- and y-component relax with the so-called transversal relaxation time, T_2. [12]

These considerations lead us to the *Bloch equations*:

$$\frac{\mathrm{d}M_x}{\mathrm{d}t} = -\Omega_S M_y - \frac{M_x}{T_2},$$

$$\frac{\mathrm{d}M_y}{\mathrm{d}t} = \Omega_S M_x - \omega_1 M_z - \frac{M_y}{T_2},$$

$$\frac{\mathrm{d}M_z}{\mathrm{d}t} = -\omega_1 M_y - \frac{M_z - M_0}{T_1} \tag{2.8}$$

2.1.4 Pulses

As mentioned above in the rotating frame the magnetization vector precesses about the z-axis with the resonance offset frequency, Ω_S. If one applies microwave (m.w.) irradiation, there is an additional precession about the m.w. field direction with the frequency ω_1. Both fields (the m.w. and static one) add up to an effective field $\mathbf{B_{eff}}$. The precession of the magnetization vector about the effective field is called *nutation*. Its frequency given by

$$\omega_{eff} = \sqrt{\Omega_S^2 + \omega_1^2}. \qquad (2.9)$$

The angle between the effective field and the z-axis is

$$\theta = arctan(\frac{\omega_1}{\Omega_S}). \qquad (2.10)$$

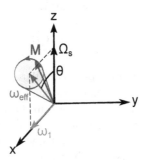

Figure 2.2: Nutation of the magnetization vector **M** during microwave irradiation with amplitude ω_1, shown in the laboratory frame. The vector moves with the frequency ω_{eff} on a cone inclined by the angle θ. [3]

If the frequency of the microwave radiation, ω_{mw}, equals the Larmor frequency, ω_0, of the observed spin, the resonance offset, Ω_S, will become zero. Thus, θ will be 90° and the nutation frequency ω_{eff} will be equal to ω_1. Such a pulse is called an *on-resonance pulse*. For an on-resonance pulse the motion of the equilibrium magnetization is very simple. It is just rotated from the z-axis towards the $-y$-axis. The angle through which the magnetization is rotated is given by:

$$\beta = w_1 * t_p, \qquad (2.11)$$

where t_p is the time for which the m.w. pulse is applied. However, if the difference between the applied m.w. radiation and the Larmor frequency is too big, then $\Omega_S \gg \omega_1$ and therefore $\theta \approx 0$ (*off-resonance pulse*) . [3, 12, 13]

Consequently, the difference between the radiation frequency and the Larmor frequency may not be too big to accomplish a good population inversion. Therefore, the inversion bandwidth of a single m.w. pulse is quite limited and we have to think of another way of inverting spins to be able to do broadband EPR. A technique for gaining high inversion efficiency over a broad bandwidth is the *fast adiabatic passage*. [14]

2.1.5 Fast Adiabatic Passage

In the fast adiabatic passage the frequency of the applied m.w. radiation varies with time. It is swept at a rate that is small compared to the m.w. amplitude (adiabatic condition). Moreover, adiabatic rotations must be much shorter than the relaxation times T_1 and T_2. If those conditions are fulfilled the magnetization vector $\mathbf{M}$ will remain parallel to the effective field $\mathbf{B_{eff}}$ during the sweep. With such an experiment, one can achieve a high excitation bandwidth with a constant flip angle even if B_1 is very inhomogeneous. [14–16]

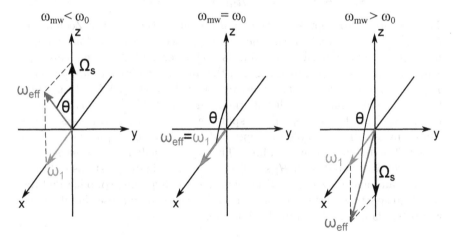

Figure 2.3: Illustration of the effective field vector, ω_{eff}, and its components, ω_1 and Ω_S during the fast adiabatic passage in a FM frame. If the adiabatic condition is fulfilled, the magnetization vector $\mathbf{M}$ will remain parallel to the effective field during the sweep. [15]

To visualize the motion of the magnetic field vector, a frequency-modulated (FM) frame is used. The FM frame rotates with the variable frequency of the applied m.w. radiation. Thus, the orientation of ω_1 remains time independent during an adiabatic passage. Figure 2.3 shows the motion of the effective field during an adiabatic pulse. When the pulse starts, $\omega_{\mathrm{mw}} < \omega_0$ therefore the effective field ω_{eff} is tilted only a bit away from the z-axis. Then the frequency increases until it is at resonance ($\omega_{\mathrm{mw}} = \omega_0$) so that ω_{eff} lies on the x-axes. After that ω_{mw} is further increased so that it becomes off resonant again ($\omega_{\mathrm{mw}} > \omega_0$). [15]

2.2 A Quantum-Mechanical Description of EPR

2.2.1 Spin-Hamiltonians

2.2.1.1 The Static Spin Hamiltonian

The static spin Hamiltonian $\mathcal{H}_0$ describes the energies of states within the ground state of a paramagnetic species. It doesn't contain space coordinates but only spin coordinates. [3]

$$\mathcal{H}_0 = \mathcal{H}_{\mathrm{EZ}} + \mathcal{H}_{\mathrm{ZFS}} + \mathcal{H}_{\mathrm{HF}} + \mathcal{H}_{\mathrm{NZ}} + \mathcal{H}_{\mathrm{NQ}} + \mathcal{H}_{\mathrm{NN}} \qquad (2.12)$$

The operator $\mathcal{H}_{\mathrm{EZ}}$ represents the Hamiltonian of the electron Zeeman interaction (see chapter 1.1.2), whereas $\mathcal{H}_{\mathrm{NZ}}$ specifies the *nuclear Zeeman interaction*. The hyperfine coupling (see chapter 1.1.3.2) is delineated by $\mathcal{H}_{\mathrm{HF}}$. The mathematical expression of the *nuclear quadrupole interaction* is $\mathcal{H}_{\mathrm{NQ}}$. This interaction is characteristic for nuclei with spin $I \geq 1$, as those have a non-spherical charge distribution. The *zero-field splitting* is described by $\mathcal{H}_{\mathrm{ZFS}}$. This field-independent splitting of the ground states is either caused by the dipole-dipole coupling of the electrons in a spin system with $S > 1/2$ or by spin-orbit coupling. The *dipole-dipole interaction between two nuclei* is covered with $\mathcal{H}_{\mathrm{NN}}$. [3] For the investigated systems only the electron and nuclear Zeeman effect as well as the hyperfine coupling and the zero-field splitting have to be considered. Therefore, only those Hamiltonians will be explained in detail in the following.

Electron Zeeman Interaction   As mentioned above an electron has an intrinsic angular momentum that is quantized in two states ($m_{\mathrm{s}} = \pm\frac{1}{2}$). The splitting of these energy levels in a magnetic field is called the Zeeman interaction. It can be described with the electron Zeeman Hamiltonian:

$$\mathcal{H}_{\mathrm{EZ}} = \frac{\beta_e}{h} \mathbf{B}_0^T \mathbf{g} \mathbf{S}, \qquad (2.13)$$

where $\mathbf{B}_0^T$ is the transpose of the static magnetic field vector and $\mathbf{g}$ is the g-tensor. [3, 17]

In anisotropic systems the Zeeman interaction can depend on the orientation of the system with respect to the external magnetic field (see chapter 1.1.3.1). Therefore, the g-factor has to be a tensor:

$$\mathbf{g} = \begin{pmatrix} g_x & 0 & 0 \\ 0 & g_y & 0 \\ 0 & 0 & g_z \end{pmatrix} \tag{2.14}$$

If $\mathbf{B}_0$ is parallel to the x(y,z)-axis of the molecule internal coordinate system, the Zeeman splitting is $\frac{\beta_e}{\hbar} B_0 g_x$ ($\frac{\beta_e}{\hbar} B_0 g_y$, $\frac{\beta_e}{\hbar} B_0 g_z$). If the external field isn't aligned with any primary axis of the molecule internal coordinate system, then all three diagonal elements contribute to the effective g-value: [3]

$$g_{\text{eff}} = \sqrt{g_x^2 \sin^2 \theta \cos^2 \varphi + g_y^2 \sin^2 \theta \sin^2 \varphi + g_z^2 \cos^2 \varphi} \tag{2.15}$$

θ and φ are the polar angles that define the orientation of the external magnetic field vector $\mathbf{B}_0$ with respect to the molecule internal coordinate system (see figure 2.4). [3]

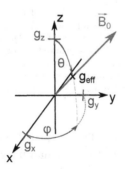

Figure 2.4: Orthorhombic model of the g-tensor shown in the laboratory frame. The coordinate systems represent the orientation of the spin with respect to the external field $\vec{B}_0$. [17]

This leads to the fact that at a constant frequency for different field strengths, molecules with different orientations are excited. Figure 2.5 shows the simulated EPR spectrum of an anisotropic system with $S = \frac{1}{2}$ and the probability that the spin with a specific orientation with respect to $\mathbf{B}_0$

is excited at a given value of B_0. The figure makes clear, that different orientations are excited at different m.w. energies or field strengths.

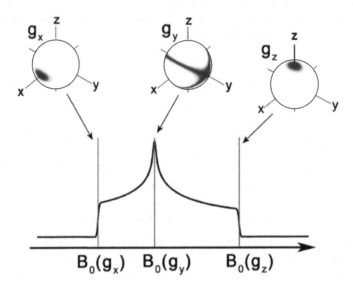

Figure 2.5: Simulation of the EPR spectrum of an anisotropic system with $S = \frac{1}{2}$. The coordinate systems represents the orientation of the spin. The x-axis of the graph shows the norm of the $\mathbf{B}_0$ vector. The spheres show the probability that the spin is excited, if the B-vector points from the point of origin to a point on the sphere. The brighter the color, the lower is the possibility that the spin is excited for this norm and orientation of $\mathbf{B}_0$. This simulation was done by Dr. Udo Kielmann, ETH Zurich, Switzerland [17].

Nuclear Zeeman Interaction   The nuclear Zeeman interaction can be expressed analogously to the electron Zeeman interaction:

$$\mathcal{H}_{\mathrm{NZ}} = \frac{\beta_{\mathrm{n}} g_{\mathrm{n}}}{h} \mathbf{B}_0^T \mathbf{I}. \tag{2.16}$$

Instead of the Bohr magneton, β_{e}, the nuclear magneton β_{n}, instead of the electron Zeeman tensor $\mathbf{g}$, the nuclear Zeeman factor g_{n} and instead of the electron spin angular momentum operator $\mathbf{S}$, the nuclear spin vector operator $\mathbf{I}$ are used. In most EPR-experiments the nuclear Zeeman interaction can be considered to be isotropic. [3]

Hyperfine Coupling The Hamiltonian of the hyperfine coupling, $\mathcal{H}_{HF}$, consists of the *Fermi contact interaction*, $\mathcal{H}_F$, and the *electron-nuclear dipole-dipole coupling*, $\mathcal{H}_{DD}$. It can be expressed as:

$$\mathcal{H}_{HF} = \mathbf{S}^T \mathbf{A} \mathbf{I}, \tag{2.17}$$

where $\mathbf{A}$ is the hyperfine tensor and $\mathbf{I}$ the nuclear spin vector operator. [3] The Fermi contact interaction accounts for the possibility that the positions of the electron and nuclear spin overlap. It is given by:

$$\mathcal{H}_F = a_{iso}\mathbf{S}^T \mathbf{I}, \tag{2.18}$$

whereat a_{iso} is the isotropic hyperfine coupling constant. While the Fermi contact interaction is isotropic, the electron-nuclear dipole-dipole coupling depends on the relative orientation of the nuclear and the electron spin. [3] It is described by

$$\mathcal{H}_{DD} = \mathbf{S}^T \mathbf{T} \mathbf{I}, \tag{2.19}$$

with $\mathbf{T}$ being the anisotropic dipolar coupling tensor. For an anisotropic g factor the hyperfine tensor $\mathbf{A}$ can now be written as: [3]

$$\mathbf{A} = a_{iso}\mathbf{1} + \frac{\mathbf{g}\mathbf{T}}{g_e}. \tag{2.20}$$

Zero-Field Splitting Strongly interacting electrons (group spin $S > \frac{1}{2}$), like for example in transition metal or lanthanide ions, are described by $\mathcal{H}_{ZFS}$:

$$\mathcal{H}_{ZFS} = \mathbf{S}^T \mathbf{D} \mathbf{S} \tag{2.21}$$

whereat $\mathbf{D}$ is the symmetric and traceless zero-field interaction tensor. In the principal axes system of $\mathbf{D}$, $\mathcal{H}_{ZFS}$ becomes:

$$\begin{aligned}
\mathcal{H}_{ZFS} &= D_x S_x^2 + D_y S_y^2 + D_z S_z^2 \\
&= D[S_z^2 - \frac{1}{3}S(S+1)] + E(S_x^2 - S_y^2).
\end{aligned} \tag{2.22}$$

$$\text{with } D = \frac{3}{2}D_z \text{ and } E = \frac{1}{2}(D_x - D_y)$$

For cubic symmetry, $D = E = 0$; for axial symmetry, $D \neq 0, E = 0$; and for even lower symmetries, $D \neq 0, E \neq 0$. For spins with $S > 2$ even further distortions have to be considered.[3]

2.2.1.2 Weak Coupling between Electron Spins

While strongly interacting electrons are only of minor importances for this thesis, the interaction between weakly coupled electron spins is the fundamental effect on which this work is based. The complete Hamiltonian for a system consisting of two weakly coupled electron spins is given by the Hamiltonians for each individual spin system $\mathcal{H}_0(S_1)$ and $\mathcal{H}_0(S_2)$ (see equation 2.12) and two coupling terms, $\mathcal{H}_{\text{exch}}$ and $\mathcal{H}_{\text{dd}}$. [3]

$$\mathcal{H}_0(S_1, S_2) = \mathcal{H}_0(S_1) + \mathcal{H}_0(S_2) + \mathcal{H}_{\text{exch}} + \mathcal{H}_{\text{dd}} \tag{2.23}$$

$\mathcal{H}_{\text{exch}}$ - **The Exchange Coupling**   The exchange coupling is caused by significant overlapping between the orbitals of the two spins. In this case the unpaired electrons can be exchanged. The Hamiltonian of the exchange coupling is given by

$$\mathcal{H}_{\text{exch}} = \mathbf{S}_1^T \mathbf{J} \mathbf{S}_2, \tag{2.24}$$

where $\mathbf{J}$ is the exchange coupling tensor. [3]

$\mathcal{H}_{\text{dd}}$ - **The Dipole-Dipole Coupling**   The operator of the dipole-dipole coupling is constructed in an analogous manner to the exchange coupling term. Instead of $\mathbf{J}$, the dipole-dipole coupling tensor $\mathbf{D}$ is used.

$$\mathcal{H}_{\text{dd}} = \mathbf{S}_1^T \mathbf{D} \mathbf{S}_2, \tag{2.25}$$

If the anisotropy of the **g**-tensor can be neglected and the two electrons can be described as point dipoles, the $\mathbf{D}$ is given by:

$$\mathbf{D} = \frac{\mu_0}{4\pi\hbar} \frac{g_1 g_2 \beta_e^2}{r_{12}^3} \begin{pmatrix} -1 & 0 & 0 \\ 0 & -1 & 0 \\ 0 & 0 & 2 \end{pmatrix} = \begin{pmatrix} -\omega_{\text{dd}} & 0 & 0 \\ 0 & -\omega_{\text{dd}} & 0 \\ 0 & 0 & 2\omega_{\text{dd}} \end{pmatrix}, \tag{2.26}$$

where r_{12} is the distance between the two electron spins, μ_0 is the vacuum permeability and ω_{dd} is the dipole-dipole coupling. [3]

2.2.1.3 The Oscillatory Hamiltonian

Transitions between different electron spin states can be induced by an electromagnetic field with a frequency close to the corresponding Larmor frequency. The interaction between the electron spin $\mathbf{S}$ and the linearly polarized oscillatory m.w. field $B_{mw}(t)$ is described by the time-independent oscillatory Hamiltonian $\mathcal{H}_1$ in the rotating frame: [3]

$$\mathcal{H}_1 = \frac{\beta_e \mathbf{B}_{mw}^T(t) \mathbf{g} \mathbf{S}}{\hbar} \tag{2.27}$$

The transition amplitude of a single-photon transition is given by

$$a_{kl} = \langle \phi_k | \mathcal{H}_1 | \phi_l \rangle, \tag{2.28}$$

whereat ϕ_k and ϕ_l are eigenstates of the static Hamiltonian $\mathcal{H}_0$. [3] The transition probability is the squared modulus of the transition amplitude: [3]

$$w_{kl} = |a_{kl}|^2. \tag{2.29}$$

2.2.2 The States of a Spin System

In the classical description of EPR we regard the interaction of the macroscopic bulk magnetization with the applied m.w. frequency. However, the magnetic moment of an electron behaves non-classical. While a classical magnetic moment can have any orientation with respect to the $\mathbf{B}_0$ field, an electron spin $S = \frac{1}{2}$ can only be observed in two different states. These states are the eigenstates α and β. The actual quantum mechanical state is represented by a superposition of α and β states. The wave function of a single spin is given by [3]

$$|\psi\rangle = c_\alpha |\alpha\rangle + c_\beta |\beta\rangle, \tag{2.30}$$

$$\text{with } |c_\alpha|^2 + |c_\beta|^2 = 1.$$

If all spins of an ensemble are in the same state, such a state is called a *pure state*. For an ensemble of isolated electron spins with $S = \frac{1}{2}$ that is in a pure state the components of the macroscopic magnetization vector are: [3]

$$M_x = \frac{-g_e \beta_e N}{2V} (c_\alpha^* c_\beta + c_\alpha c_\beta^*) \tag{2.31}$$

$$M_y = \frac{i g_e \beta_e N}{2V} (c_\alpha^* c_\beta - c_\alpha c_\beta^*) \tag{2.32}$$

$$M_z = \frac{-g_e \beta_e N}{2V} (|c_\alpha|^2 - |c_\beta|^2) \tag{2.33}$$

Whereat V is the ensemble volume. Any ensemble of spins can be described by a superposition of n sub ensembles ($n \leq N$) in pure states. The corresponding state is called a *mixed state*. The magnetization components are given by the average over the coherent sub ensembles: [3]

$$M_x = \frac{-g_e \beta_e N}{2V} \sum_i p_i (c_\alpha^{(i)*} c_\beta^{(i)} + c_\alpha^{(i)} c_\beta^{(i)*}) \qquad (2.34)$$

$$M_y = \frac{i g_e \beta_e N}{2V} \sum_i p_i (c_\alpha^{(i)*} c_\beta^{(i)} - c_\alpha^{(i)} c_\beta^{(i)*}) \qquad (2.35)$$

$$M_z = \frac{-g_e \beta_e N}{2V} \sum_i p_i (|c_\alpha^{(i)}|^2 - |c_\beta^{(i)}|^2) \qquad (2.36)$$

$$\text{with } \sum_i^n p_i = 1$$

By measuring M_x, M_y and M_z we can obtain only the averages of the coefficients: $\sum_i p_i |c_\alpha^{(i)}|^2 = \overline{|c_\alpha|^2}$, $\sum_i p_i c_\alpha^{(i)} c_\beta^{(i)*} = \overline{c_\alpha c_\beta^*}$, $\sum_i p_i c_\alpha^{(i)*} c_\beta^{(i)} = \overline{c_\alpha^* c_\beta}$ and $\sum_i p_i |c_\beta^{(i)}|^2 = \overline{|c_\beta|^2}$. These ensemble averages can be arranged in a *density matrix*: [3]

$$\sigma = \begin{pmatrix} \overline{|c_\alpha|^2} & \overline{c_\alpha c_\beta^*} \\ \overline{c_\alpha^* c_\beta} & \overline{|c_\beta|^2} \end{pmatrix} \qquad (2.37)$$

$$\text{with } \overline{|c_\alpha|^2} + \overline{|c_\beta|^2} = 1.$$

The diagonal elements of the density matrix are the populations of the basis states. The difference between the diagonal elements thus represents the *polarization*. The off-diagonal elements represent *coherences*. [3] The term "coherence" denotes here "phase coherence" between spin state functions. In general the density matrix for a state ψ with n energy eigenstates $|k\rangle$ is a $n \times n$ Hermitian. Hence, the density matrix of a $S = 7/2$ system is a 8×8 Hermitian. The ij's element of this matrix is the statistically adequate average over the product of the expansion coefficients $\overline{c_i c_j^*}$, with $|\psi\rangle = \sum_{k=1}^n c_k |k\rangle$. The matrix trace, $\sum_{k=1}^n |c_k|^2$, is unity. There is phase coherence between specific pairs of eigenstates, if the corresponding off-diagonal elements of the density matrix are non-zero. One can distinguish between different types of coherences by the difference in the magnetic quantum number, Δm_S, of the corresponding states. Thus, it is more appropriate to use the term 'x-quantum coherence' (with $x = |\Delta m_S| = $

$0, 1, 2, ...$). As mentioned above, for a transition between the Zeeman levels the conservation of angular momentum imposes a selection rule of $|\Delta m_s| = 1$. Therefore, only single-quantum coherences can be detected by magnetic resonance. Consequently, single-quantum coherences imply the presence of transverse magnetization. Coherence transfer between spin states can be accomplished by the use of appropriate $\mathbf{B}_1$ pulses. [2]

2.2.3 Time Evolution of a Spin System

The evolution of a spin system can be described by a quantum mechanical analog of the relaxation-free Bloch equations, the *Liouville-von Neumann equation*:

$$\boxed{\frac{d\sigma}{dt} = -i[\mathcal{H}(t), \sigma(t)]}$$

(2.38)

If $\mathcal{H}$ is time-independent, the integration of equation 2.38 yields:

$$\sigma(t) = e^{-i\mathcal{H}t}\sigma(0)e^{i\mathcal{H}t}$$

(2.39)

with $U(t) = e^{-i\mathcal{H}t}$ and accordingly $U^{-1}(t) = U^{\dagger}(t) = e^{i\mathcal{H}t}$

$U(t)$ is called a *propagator*, as it propagates the density operator in time. [3] The propagator method is used for numerical calculations. Therefore, the evololution of the spin system is subdivided into short time intervalls during which $\mathcal{H}$ can be considered to be time-independent. [3]

2.3 Important Pulse EPR Experiments

2.3.1 Echo Detected Field Sweep (ESE)

To understand the echo-detected field sweep (ESE) experiment one first has to take a look at the echo experiment which is described in detail in figure 2.6.

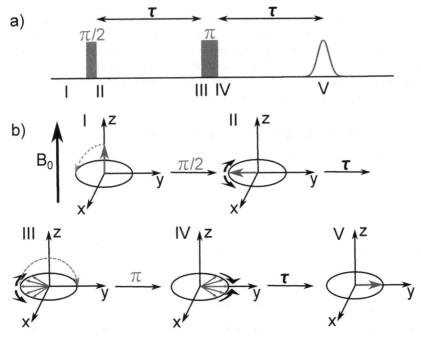

Figure 2.6: Sketch of the echo experiment. a) Pulse sequence of the echo experiment.
b) Illustration of the movement of the magnetization vector during the
echo experiment in the rotating frame: I: Before the first pulse is applied
the magnetization vector is aligned with the external magnetic field. II:
The first $\pi/2$ pulse causes a flip in -y-direction. III: Defocusing of the spins
over a time τ. IV: Refocusing with an inversion π pulse. V: Detection of
the echo signal after second time τ. [17]

The pulse sequence of the echo experiment consists of a $\pi/2$ pulse followed
by a π pulse. The $\pi/2$ pulse rotates the bulk magnetization away from its
equilibrium position parallel to the external magnetic field into the xy-plane.
As a spins with different resonance offsets are flipped, during an evolution
time τ the spins start to defocus due to little differences in their Larmor
frequencies. The following π pulse flips the spins by 180 degree around the
x-axis. Because of this flip, the rotation of the spins in the rotating frame is
reversed and leads to a refocusing. After an evolution time τ all spins are
aligned parallel to the y-axis and the signal is detected. This method is very
important for EPR, as the inhomogeneous broadening can be overcome by
the refocusing π pulse. [17]

In the ESE experiment, the intensity of the echo signal is detected as a function of the magnetic field.

2.3.2 Determination of the Nutation Frequency

As mentioned above the precession of the magnetization vector about the effective field is called nutation (see chapter 2.1.4). The nutation frequency, ω_{eff}, depends on the frequency of the applied pulse, its amplitude and the Larmor frequency of the observed spins.

To calculate the flip angle of an applied m.w. pulse, one has to know its nutation frequency (see equation 2.11). There are various methods to determine the nutation frequency. Figure 2.7 shows the experiment used in this thesis. First a nutation pulse of variable length t_1 is applied. This pulse rotates the bulk magnetization away from its equilibrium position parallel to the z-axes. After a delay time T, an echo pulse sequence is applied to detect the z-part of the magnetization vector (longitudinal magnetization). If the flip angle of the nutation pulse was 90 degree, the value of the detected longitudinal magnetization is zero. If it was 180 degree the detected magnetization has the negative value of the equilibrium magnetization and so on. This experiment is repeated various times. The nutation frequency of the pulse can be determined from the oscillation of the longitudinal magnetization.

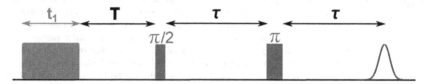

Figure 2.7: Echo detection of the longitudinal magnetization after a nutation pulse. [3]

2.3.3 Double Electron-Electron Resonance (DEER)

2.3.3.1 A General Description of the DEER Experiment

The four-pulse DEER sequence (see figure 2.8) is the most popular technique for distance measurements between spin labels (spin 1 and spin 2 in figure 2.9) in the range of approximately 1.8 to 6 nm. With the DEER technique one has direct access to the distance-dependent pairwise dipole-dipole coupling between electron spins ω_{dd}. [18]

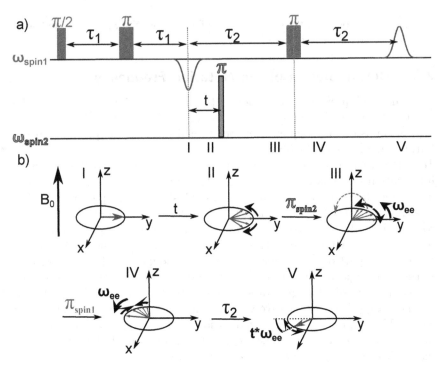

Figure 2.8: Four-pulse DEER experiment. a) Pulse sequence. b) Illustration of the
movement of the magnetization vector in the vector picture (rotating
frame). I: After the initial echo sequence the spins are refocused on the
$+y$-axes. II: Defocusing of the spins over a time t. III: The π pump pulse
at frequency ω_{spin2} inverts the state of spin 2. Therefore, its local fields
change their direction and the effective field at spin 1 changes. Thus, the
precession frequency of spin 1 changes and an additional rotation with
the frequency ω_{ee} appears in the rotating frame. IV: Refocusing with an
inversion π pulse. V: After second time τ_2 the signal (=DEER echo) is
detected. [18, 19]

The DEER experiment starts with an echo sequence at the observer frequency
ω_{spin1}. After this the observed spins (spin 1 in figure 2.9) evolve for a time
period τ_2. During this period at a variable time t a π pulse at a different
frequency ω_2 is applied. This pump pulse inverts the state of spin 2 (see
figure 2.9). Due to this the local fields caused by spin 2 are inverted,
too. Therefore, the effective field at spin 1 changes. Thus, the precession
frequency of spin 1 changes and an additional rotation with the frequency
of the electron-electron coupling ω_{ee} appears in the rotating frame. When

time period τ_2 is terminated, a second π pulse at the observer frequency ω_1 is applied to refocus the spins 1. At the time of detection the magnetization is then out of phase by the angle

$$\Delta\phi_{ee} = t * \omega_{ee}. \tag{2.40}$$

Hence, ω_{ee} can be determined by observing the amplitude of the DEER signal as a function of t. [3, 18]

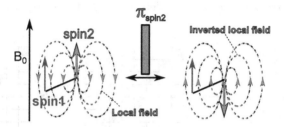

Figure 2.9: Local field picture of the DEER experiment. The π pump pulse at frequency ω_{spin2} inverts the state of spin 2. Therefore, the direction of the local fields caused by spin 2 changes. [19]

In the point-dipole approximation the electron-electron coupling ω_{ee} can be calculated from the dipole-dipole coupling ω_{dd} and the exchange coupling J (see chapter 2.2.1.2):

$$\omega_{ee} = \omega_{dd} + J = \frac{\mu_0}{4\pi\hbar} \frac{g_1 g_2 \beta_e^2}{r_{12}^3} (3\cos^2\theta_{dd} - 1) + J, \tag{2.41}$$

whereat θ_{dd} is the angle between the inter-spin vector and the external magnetic field $\mathbf{B}_0$. The exchange coupling J is assumed to be negligible small for inter-spin distances larger than 2 nm. Thus, the distance dependent dipole-dipole coupling ω_{dd} can be determined directly. [3]

2.3.3.2 Echo Reduction in High Spin Systems

If high-spin systems with $S > 1/2$ are used as spin labels, the vector picture is not an appropriate description anymore. We have to look directly at the density matrix of the observed spin. Moreover, due to the non-observed transitions, unwanted effects may occur. Of those the DEER echo reduction is of significant importance for this thesis.

The echo reduction effect has been observed for Gd(III)-nitroxide DEER. It causes a decrease in intensity of the Gd(III) echo in the presence of the nitroxide pump pulse. [6] In 2012 *Yulikov, M. et al.* proposed the mechanism for the DEER echo reduction as follows. If there is an overlap between the absorption spectra of the Gd(III) and the nitroxide, then the pump pulse may hit not only the $| -1/2 \rangle \leftrightarrow | +1/2 \rangle$ transition of the nitroxide pump spin, but also a transition of the Gd(III). In the case that this transition has a level in common with the observed transition, part of the coherence is transfered to other elements of the density matrix (see figure 2.10). Due to the loss of the phase coherence between the to energy levels $| -1/2 \rangle$ and $| +1/2 \rangle$ this transition cannot be detected later in the experiment. [20]

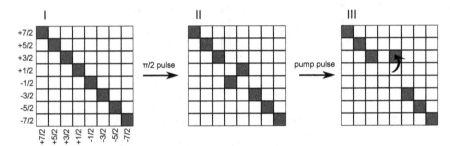

Figure 2.10: Mechanism of the echo reduction in a DEER experiment on a observer spin with $S = 7/2$. Shown is the qualitative density matrix of a $S = 7/2$ system. Every square belongs to one matrix element. The filling indicates matrix elements that are non-zero, while the squares without a filling stand for zero matrix elements. In the discussed DEER experiment, the observer frequency is set at the $| -1/2 \rangle \leftrightarrow | +1/2 \rangle$ transition of the $S = 7/2$ spin system. I: Before the first pulse is applied the spins are quantized along the external magnetic field. The diagonal elements are non-zero. II: The $\pi/2$ observation pulse (at ω_{spin1}) generates single-quantum coherence between the $| -1/2 \rangle$ and $| +1/2 \rangle$ eigenstates. III: The pump pulse (at ω_{spin2}), that should only affect the pumped spins, leads to unwanted coherence transfer from the observed transition to other elements of the density matrix. [20]

2.3.3.3 The Pake Pattern

The amplitude of the DEER signal of a system with one observer spin and one pumped spin (two-spin system) is:

$$V_{\text{pair}}(t) = B(t) \int_0^{\frac{\pi}{2}} V_{\text{pair}}(t, \theta_{\text{dd}}, r_{12}) \sin\theta_{\text{dd}} d\theta_{\text{dd}} \qquad (2.42)$$

with $V_{\text{pair}}(t, \theta_{\text{dd}}, r_{12}) = 1 - \lambda[1 - \cos(\omega_{\text{dd}}(\theta_{\text{dd}}, r_{12})t)]$, $\qquad (2.43)$

whereat the background factor $B(t)$ takes into account all intermolecular contributions to the DEER signal and λ is a parameter that describes the fraction of pumped spins. For a homogeneous distribution of nanoobjects $B(t)$ can be fitted with

$$B(t) = exp(-k_{\text{dec}}t^{d/3}), \qquad (2.44)$$

where d is the dimensionality of the homogeneous distribution and k_{dec} is the rate constant of the decay of the intermolecular couplings. [1, 21]. Figure 2.11 shows the spectrum which is obtained by the Fourier transformation of $V_{\text{pair}}(t)$, if every possible orientation of the inter-spin vector with respect to the external magnetic field $\mathbf{B}_0$ contributes to the spectrum. Such dipolar coupling spectra are called *Pake patterns*. [3]

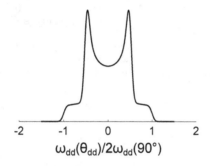

Figure 2.11: Schematic Pake pattern.

The Pake pattern is formed by adding up the dipole-dipole couplings ω_{dd} of all possible orientations of the inter-spin vector with respect to the external magnetic field $\mathbf{B}_0$ with a $\sin(\theta_{\text{dd}})$ weighting (see figure 2.12). The reason for the $\sin(\theta_{\text{dd}})$ weighting is that the amount of possible orientations of the inter-spin vector per θ_{dd} value is proportional to $\sin(\theta_{\text{dd}})$ (see figure 2.13). [19, 22, 23]

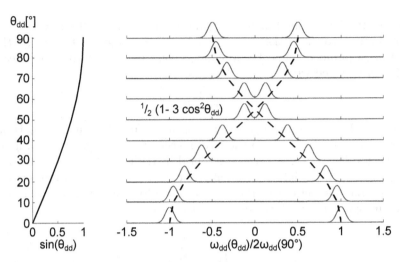

Figure 2.12: Origin of the Pake pattern: right: ω_{dd} spectra in dependency of θ_{dd}, left: weighting of each ω_{dd} spectrum. The Pake pattern is formed by multiplying every individual $\omega_{dd}(\theta_{dd})$ spectrum with $\sin(\theta_{dd})$ and adding them up. [19]

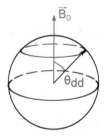

Figure 2.13: Weighting of the contribution of a specific ω_{dd} to the whole pattern as a function of θ_{dd}. θ_{dd} is the angle between the inter-spin vector and the external magnetic field $\mathbf{B}_0$. If the inter-spin vector points from the point of origin to a point on the sphere, the blue circle on the sphere shows all possible orientations of the inter-spin vector with respect to $\mathbf{B}_0$ for a specific value of θ_{dd}. The circumference of the blue circle is $2\pi \sin(\theta_{dd})r_{12}$. Therefore, the contribution of a specific $\omega_{dd}(\theta_{dd})$ to the entire DEER signal is proportional to $\sin(\theta_{dd})$. [19]

2.3.3.4 Orientation Selection in DEER

For DEER it is only important that every possible orientations of the inter-spin vector with respect to the external magnetic field $\mathbf{B_0}$ contributes to the signal. It doesn't matter if specific spin orientations are selected. If at a given observer position a set of orientations of spin 1 is selected and the orientation of spin 2 is not correlated to the orientation of spin 1, there is no orientation selection in the DEER signal. This is due to the fact that, for every orientation of the inter-spin vector, there is the same amount of possible orientations of spins 2 connected to the selected spins 1 that can be excited by the pump pulse. Every inter-spin vector has therefore the same probability that it may contribute to the DEER signal.

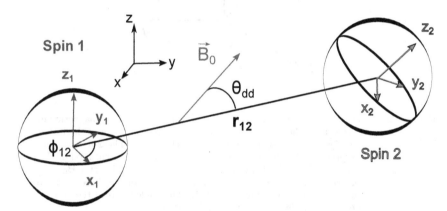

Figure 2.14: Model of a geometrically correlated spin pair: Pair geometry is defined with respect to the molecular frame of spin 1 (x_1, y_1, z_1) by six parameters: the interspin distance r_{12}, the polar angles θ_{dd}, Φ_{12} specifying orientation of the interspin axes in molecular frame of spin 1 and the Euler angles α, β and γ defining relative orientation of the frame of spin 2 (x_2, y_2, z_2) with respect to the frame of spin 1. For numerical computations all frames are defined with respect to a common reference frame (x, y, z). [21]

Figure 2.14 shows the model of a geometrically correlated spin pair. As explained above, at a given observer position, a set of orientations of spin 1 is selected. However, for a geometrically correlated spin pair, the selection of spins 1 by observer orientation selection corresponds to a selection of orientations of the molecular frame of the spins 2. [21] Due to the preselection of spins 2, the case may occur that for specific θ_{dd} values no spins 2 can be excited. Then some possible orientations of the inter-spin vector with respect to the external magnetic field $\mathbf{B}_0$ do not contribute to the spectrum. As a consequence the dipolar coupling spectra will not be a Pake pattern.

2.3.3.5 The Three-Spin Contribution in DEER

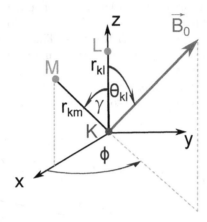

Figure 2.15: Orientation of a triangular system with respect to the magnetic field vector $\mathbf{B}_0$. The observer spin K is placed at the origin of the frame. The first pumped spin L is placed on the z-axis. The x-axis is defined by placing the third spin M in the xz-plane. The orientation of the three-spin system with respect to the magnetic field vector $\mathbf{B}_0$ is described by the polar angles θ_{kl} and ϕ. Each of the three spins can take the role of the observer spin K. The angle between the two observer spin-pump spin vectors is γ [1, 10]

Figure 2.15 defines all parameters that are needed to describe the orientation of a three-spin system with respect to the magnetic field vector $\mathbf{B}_0$. If all spins of a three-spin system have the same EPR spectrum, each of them can take the role of observer or pumped spin. Therefore, we have to consider all three inter-spin vectors in a three-spin system. The DEER signal $V(t)$ is then obtained by summing over the signal of all three observer spins, which consists of the product of the two pairwise signals: [1, 10]

$$V_3(t) = B(t)\frac{1}{3}\sum_{k=1}^{3}\prod_{l\neq k}^{3} f_{kl}(t,\theta_{kl},r_{kl}) \qquad (2.45)$$

$f_{kl}(t,\theta_{kl},r_{kl})$ is the dipolar evolution function of the spin pair (k,l).

$$f_{kl}(t,\theta_{kl},r_{kl}) = 1 - \lambda_l[1 - \cos(\omega_{kl}(\theta_{kl},r_{kl})t)] \qquad (2.46)$$

By correction of the inter-molecular background, we obtain the form factor for a particular orientation relative to the magnetic field vector $\mathbf{B}_0$: [1, 10]

$$F_3(t, \theta, \phi) = V_3(t)/B(t) = \frac{1}{3} \sum_{k=1}^{3} \prod_{l \neq k}^{3} f_{kl}(t, \theta_{kl}) \tag{2.47}$$

If only the K spin took the role of the observer spin, the form factor for one particular orientation could be expressed as:

$$\begin{aligned} F_3(t, \theta, \phi) =& 1 - \lambda_m - \lambda_l + \lambda_m \lambda_l + \\ & (\lambda_l - \lambda_m \lambda_l) \cos(\omega_{kl} t) + (\lambda_m - \lambda_m \lambda_l) \cos(\omega_{km} t) + \\ & \lambda_m \lambda_l \cos(\omega_{km} t) \cos(\omega_{kl} t) \end{aligned} \tag{2.48}$$

The dipolar frequencies ω_{kl} and ω_{km} depend on the inter-spin distances r_{kl} and accordingly r_{km} and on the angles θ_{kl} respectively θ_{km}:

$$\omega_{kl} = \frac{\mu_0}{4\pi\hbar} \frac{g_K g_L \beta_e^2}{r_{kl}^3} (1 - 3\cos^2\theta_{kl}) \tag{2.49}$$

$$\omega_{km} = \frac{\mu_0}{4\pi\hbar} \frac{g_K g_M \beta_e^2}{r_{km}^3} (1 - 3\cos^2\theta_{km}) \tag{2.50}$$

g_K, g_M and g_L are the effective g-factors of the three spins. The third frequency ω_{ml} is encountered when L or M takes the role of the observer spin. [10] If one takes a closer look at equation 2.48, three different types of contributions to the form factor can be identified. The frequency independent part depends only on the inversion efficencies of the pumped spins, λ_l and λ_m. If the labelling efficiency is 100 %, the modulation depth[1] can be defined as 1 minus this frequency independent part:

$$\Delta = 1 - (1 - \lambda_m - \lambda_l + \lambda_m \lambda_l) \tag{2.51}$$

The second part that depends on just one frequency is called the two-spin contribution, $P(t)$. The two-spin contribution for one particular orientation and K as observer spin is given as:

$$P(t, \theta, \phi) = (\lambda_l - \lambda_m \lambda_l) \cos(\omega_{kl} t) + (\lambda_m - \lambda_m \lambda_l) \cos(\omega_{km} t) \tag{2.52}$$

---

[1]The modulation depth Δ is defined as one minus the value of the phase and background corrected DEER trace in its constant region. It is directly related to the percentage of spins that have been flipped. Δ is always in the range from 0 to 1.

The third contribution to the form factor is the three-spin contribution, $T(t)$.

$$T(t, \theta, \phi) = \lambda_m \lambda_l \cos(\omega_{km} t) \cos(\omega_{kl} t) \tag{2.53}$$

The product of two cosine functions can be rewritten as the sum of cosines of the combination and difference frequency:

$$cos(\omega_{km} t) \cos(\omega_{kl} t) = \frac{1}{2} \{\cos[(\omega_{kl} + \omega_{km}) t] + \cos[(\omega_{kl} - \omega_{km}) t]\}, \tag{2.54}$$

under the assumption of a two-spin system, the three-spin contribution will lead to false inter-spin distances during the conversion from the time to the distance domain. [1, 10] Otherwise $T(t)$ is also an additional source of information.

Up until now we discussed the contributions to the form factor for one particular orientation. As our experiments were performed on glassy frozen solutions and not on single crystals, it is impossible to get the form factor for only one orientation. Since nanoobjects will always be distributed uniformly in powders or glassy frozen solutions, we have to integrate over all possible orientations to compute the form factor that we can extract from our measurements: [10]

$$F_{powder}(t) = \frac{1}{2\pi} \int_0^{2\pi} \int_0^{\pi/2} F_3(t, \theta, \phi) \sin(\theta) d\theta d\phi \tag{2.55}$$

If we do the same for the two-spin contribution, we obtain:

$$P_{powder}(t) = \frac{1}{2\pi} \int_0^{2\pi} \int_0^{\pi/2} P(t, \theta, \phi) \sin(\theta) d\theta d\phi = \int_0^{\pi/2} P(t, \theta) \sin(\theta) d\theta \tag{2.56}$$

By powder averaging the dependency on ϕ is lost for the two-spin contribution. [1] This is due to the fact that $P(t, \theta, \phi)$ consists simply of a sum over cosine functions of different dipolar frequencies. The powder average can thus be performed over each summand separately, which integrates the ϕ dependency out. Consequently, we do not have to care about the relation between θ_{kl} and θ_{km} and can look at them independently. Needless to say the situation is different for the three-spin contribution. As we have to deal with the product of cosine functions of different dipolar frequencies, we cannot split the integrals easily:

$$T_{powder}(t) = \frac{1}{2\pi} \int_0^{2\pi} \int_0^{\pi/2} T(t, \theta, \phi) \sin(\theta) d\theta d\phi \qquad (2.57)$$

As a consequence, we have to consider the relation between θ_{kl} and θ_{km}. The angle θ_{km} can be expressed as: [1, 10]

$$\cos(\theta_{km}) = \sin(\gamma) \cos(\theta_{kl}) \cos(\phi) + \cos(\gamma) \cos(\theta_{kl}) \qquad (2.58)$$

Therefore, the three-spin contribution depends on the angle γ between the two inter-spin vectors $\overrightarrow{KL}$ and $\overrightarrow{KM}$, while the two-spin contribution is independent from this angle.

2.4 The Investigated Molecules

2.4.1 Nitroxide Radicals

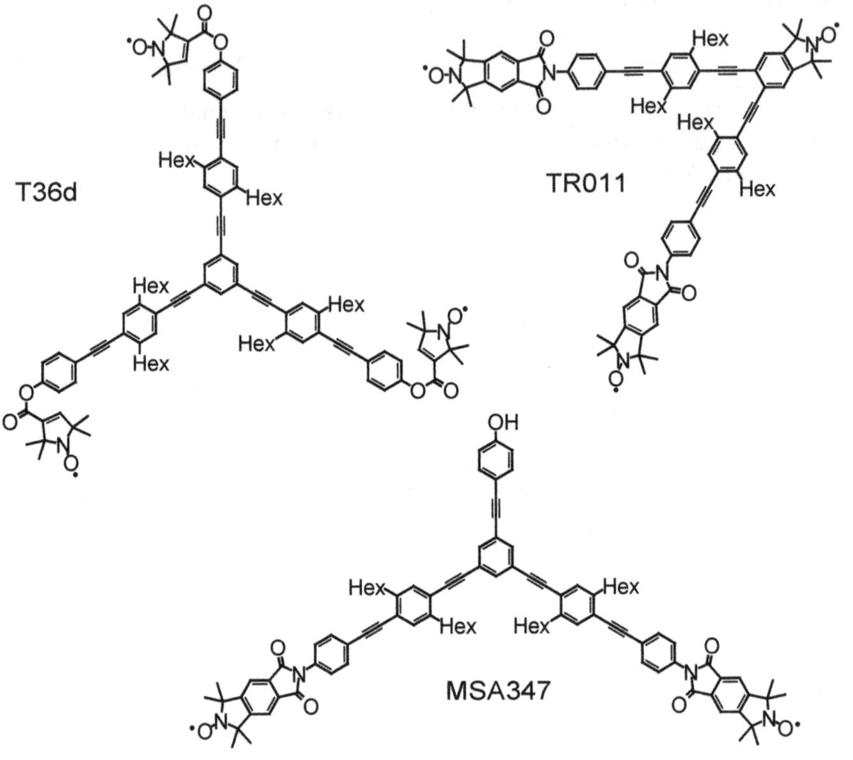

Figure 2.16: The investigated nitroxide radicals: The model compounds: the T36d triradical plus the TR011 triradical and the reference compound: the MSA347 diradical

The investigated nitroxide radicals (figure 2.16) were synthesized in the group of Prof. Adelheid Godt, Bielefeld University, Germany. They were prepared to fulfill two special criteria. First the variations in the distances and angles between the electron spins due to conformational flexibility should be minimized. Second the amount of spin density that is coupled into the polyconjugated backbone should be reduced to a minimum. [1] Unpaired electrons were introduced using nitroxide spin labels. For the triradical T36d, 1-oxyl-2,2,5,5-tetramethylpyrroline-3-carboxylic acid was used as a spin label.

As the N-O is bond is not collinear with the carbon-backbone, rotation around the C-C triple bond results in a limited but detectable contribution to the distribution of distances and angles. For the TR011 triradical and the MSA347 diradical, 1,1,3,3-tetramethyl-2-oxylisoindole-5,6-dicarboximide was used as a spin label. This spin label ensures that the N-O bond is collinear with the carbon-backbone. Therefore, rotation around the C-C triple bond do not influence the distribution of distances and angles. [1] For ^{14}N the spin quantum number I is 1 and the dimensionless nuclear g-factor is 0.40376100. Therefore, the EPR spectra of nitroxide radicals are composed of three sub-spectra, one for each nuclear spin state ($m_I = -1, 0, +1$). Each of this sub-spectra is anisotropically broadend. [24] Figure 2.17 shows simulations of the nitroxide sub-spectra for X, Q and W band.

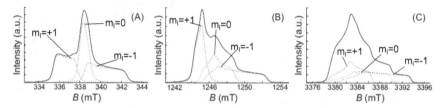

Figure 2.17: Simulations of EPR spectra of nitroxide radicals by Dr. Maxim Yulikov, ETH Zurich, Switzerland. Simulation parameters: g-tensor eigenvalues - [2.0085 2.0061 2.0022], hyperfine tensor eigenvalues - [13 13 100] MHz. (A) X band spectra with a detection frequency of 9.5 GHz. (B) Q band spectra with a detection frequency of 35 GHz. (C) W band spectra with a detection frequency of 95 GHz. [24]

2.4.2 The Gd(III)-Nitroxide Radical

The investigated Gd(III)-nitroxide radical is a poly(proline) (18-mer) with an N-terminal Gd(III)-DOTA complex label (S = 7/2) and nitroxide labels (S = 1/2) at residues 5 and 11 (see figure 2.18). The poly(proline) was synthesized in the group of Prof. Helma Wennemers, ETH Zurich, Switzerland. As the $| -1/2\rangle \leftrightarrow | +1/2\rangle$ transition of Gd(III) does not spectroscopically overlap with the nitroxide spectrum at Q band, both can be detected almost independently from each other (they are ßpectroscopically orthogonal"). In addition Gd(III) centers show much less pronounced orientation selection than nitroxide labels. As Gd(III) is a S = 7/2 spin system it shows zero-field splitting, one would expect a strong distortion of the dipolar frequencies. However, for Gd(III) in chelate complexes and for the typical experimental settings of a Gd(III)-nitroxide DEER this effect is very weak. [25–27]

This system thus allows the selective detection of the DEER trace of just one observer spin, Gd(III). Moreover, orientation selection due to the position of the observation frequency can be excluded.

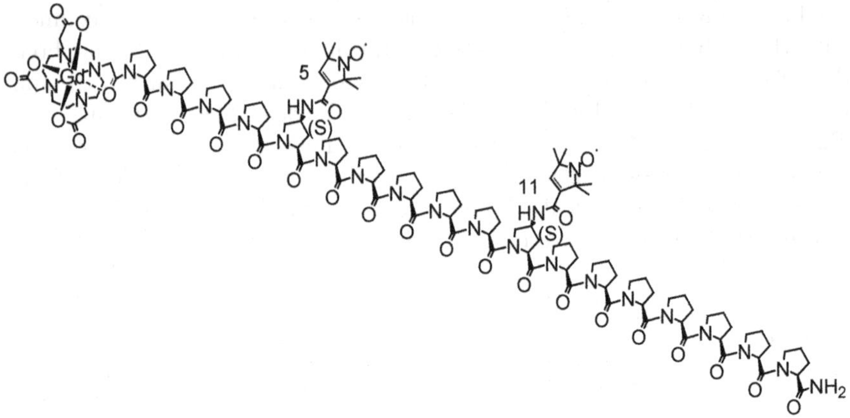

Figure 2.18: The investigated Gd(III)-nitroxide radical: A triply labeled poly(proline) (18-mer) with an N-terminal Gd(III)-DOTA complex label and nitroxide labels at residues 5 and 11.

2.5 Instrumentation

2.5.1 Basic Setup of a Pulse EPR Spectrometer

In a typical pulse EPR spectrometer a m.w. pulse forming unit (MPFU) creates pulses out of the continuous microwaves from the m.w. source. After creation the m.w. pulses are amplified in a traveling wave tube amplifier. The exact pulse power is adjusted by a precision attenuator at the output of the power amplifier. The pulses enter the resonator via the circulator. [28] The sample itself is placed in the resonator. The resonator has two main functions. First, it concentrates the magnetic component of the m.w. radiation in the sample in a direction perpendicular to B_0 and second the resonance process improves the coupling to the spins (see next section). This task is accomplished by designing the resonator such that specific frequencies form stationary waves. The concentration of the magnetic component is important as the spins interact with this component. The electric part of the radiation has to be concentrated in sample-free regions to reduce the non-resonant loss of intensity by interactions between the solvent and the electric field. [12]

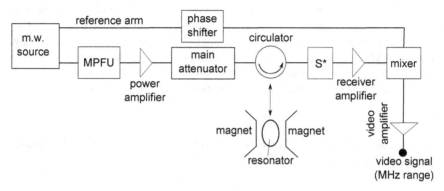

Figure 2.19: Schematic drawing of a pulse EPR spectrometer: * S stands for protection switch or power limiter. [28]

During the excitation of the sample, the receiver amplifier is protected from reflected m.w. pulses by a protection switch or power limiter. [28] After turning off the m.w. source, the spin system becomes the microwave source. [29] The signal is then amplified by the receiver amplifier and fed to a mixer, where a video signal at the difference frequency between the reference m.w. and the signal is created. This frequency subtraction corresponds to detection in the rotating frame, as only the phase difference ϕ_{ee} (see equation 2.40) is detected. To extract the sign information of the difference frequency, two mixers with a phase difference in the reference frequency of 90° are used. This corresponds to two measured signals with a phase difference of 90° in the rotating frame. They are referred to as *real*, $\cos(\omega_{dd}t)$, and *imaginary*, $\sin(\omega_{dd}t)$, signal. Measuring a signal in such a way is called *quadrature detection*. It is convenient to combine the magnetization components to the complex signal function:

$$s(t) = \cos(\omega_{dd}t) + i\sin(\omega_{dd}t) = exp(i\omega_{dd}t) \qquad (2.59)$$

The video signal is then passed through a low pass filter, amplified and detected by a fast digitizer. [28]

2.5.2 Coupling of the Resonator

The m.w. radiation is transmitted to the resonator via a waveguide. In general maximum power is transfered into a load at the end of a transmission line when the load *impedance*[2] is equal to the characteristic impedance of the line. Thus, for best sensitivity the impedance of the waveguide, Z_0,(typically about 50 Ω.) has to be matched to the impedance of the cavity resonator, Z. This is achieved by changing the coupling n between the waveguide and the resonator until the condition for *critical coupling* is fulfilled:

$$Z_0 n^2 = Z \tag{2.63}$$

Tuning and matching have to be repeated for each new sample. The impedance of the resonator is given by:

$$Z = R + i(\omega L - \frac{1}{\omega C}), \tag{2.64}$$

where R is the resistance, L is the inductance and C is the capacitance of the resonator. For a cavity resonator R is due to all power losses in the cavity, L originates from the energy stored in the magnetic field and C is caused by the electric field energy. The resonance frequency of the resonator is determined by L and C. At this frequency, the resonator appears to the transmission line to be a pure resistance. For critical coupling, this resistance value is the same as that of the transmission line. To quantify the coupling, one can define a *coupling coefficient* ε:

$$\varepsilon = \frac{Z_0 n^2}{R} \tag{2.65}$$

---

[2]Impedance extends the concept of resistance to an alternate current (AC) circuit. For an AC or an electromagnetic field, voltage and current are time depend:

$$V_t(t) = V_0 \cos(\omega t - \phi) = Re(V_0 e^{-i\phi} e^{i\omega t}) = Re(V e^{i\omega t}), \tag{2.60}$$

$$I_t(t) = I_0 \cos(\omega t - \psi) = Re(I_0 e^{-i\psi} e^{i\omega t}) = Re(I e^{i\omega t}), \tag{2.61}$$

whereat V and I are the complex voltage and current amplitudes and ϕ and ψ are phases, which generally differ for voltage and current. The impedance is defined as the complex ratio of the voltage to the current in an AC circuit:

$$Z = \frac{V}{I} \tag{2.62}$$

The imaginary part of the impedance is the *reactance*, the real part the resistance. [30]

For critical coupling $\varepsilon = 1$. The *loaded quality factor*, which characterizes the resonant enhancement of the m.w. field in the cavity, is given at resonance by:

$$Q_L = \frac{2\pi\nu_{resonator}L}{Z_0 n^2 + R} = \frac{2\pi\nu_{resonator}L}{(1+\varepsilon)R}, \qquad (2.66)$$

whereat $\nu_{resonator}$ is the resonance frequency of the resonator. [28, 29]
The amplitude of the magnetic field component B_1 in the resonator is directly proportional to the square root of the quality factor:

$$B_1 = \sqrt{\frac{2Q_L P_{mw}}{\mu_0 V_c \nu_{resonator}}}, \qquad (2.67)$$

where V_c is the effective volume of the resonator and P_{mw} the m.w. power. The bandwidth $\Delta\nu$ of the resonator is however indirectly proportional to Q_L:

$$\Delta\nu = \frac{\nu_{resonator}}{Q_L} \qquad (2.68)$$

For pulsed EPR, broad bandwidths are required. This can be achieved either by intentionally spoiling the resonator quality (increasing Z) or by *overcoupling* (increasing n beyond critical coupling, so that $\varepsilon > 1$). [28, 29] In 1981 *Mims and Peisach et al.* stated that it is better to achieve a low Q_L value by overcoupling than by introducing loss into the resonator. [31]
For $\varepsilon \neq 1$, there is a mismatch of impedance between the waveguide and the resonator. As a result, an incoming voltage wave will be reflected with a voltage-reflection coefficient given by:

$$K_{\nu\varepsilon} = \frac{1-\varepsilon}{1+\varepsilon} \qquad (2.69)$$

The power is proportional to the square of the voltage. Thus, the power-reflection coefficient is given by:

$$K_{P\varepsilon} = K_{\nu\varepsilon}^2 = \frac{(1-\varepsilon)^2}{(1+\varepsilon)^2} \qquad (2.70)$$

Besides these power losses, the decrease in Q_L leads to a smaller mw amplitude B_1 at a given power P_{mw} and to a loss in sensitivity. To compensate for these unwanted effects, pulse EPR resonators are optimized for high filling factors η.[3] [28, 29]

2.5.3 The Echo Signal for the Overcoupled Case

For a given sample the echo signal generated by the same pulse sequence at the same magnetic field strength has the same shape, regardless of the Q_L value (only true if the EPR signal width is small compared to the resonator bandwidth). The amplitude of the echo signal however depends on the coupling: [29]

$$V_{E\varepsilon} = \frac{\varepsilon}{1+\varepsilon}\sqrt{\frac{Z_0}{\varepsilon R}}V_{E},\qquad(2.72)$$

whereat $V_{E\varepsilon}$ is the echo signal at the detector for a coupling coefficient ε and V_E is the echo voltage induced in the resonator of inductance L. To normalize equation 2.72 it is convenient to divide it by the echo signal at the detector for the critically coupled case, V_{E0}: [29]

$$\frac{V_{E\varepsilon}}{V_{E0}} = \frac{2\sqrt{\varepsilon}}{1+\varepsilon}\qquad(2.73)$$

---

[3]The filling factor is the ratio of the integrals of the m.w. magnetic field amplitude over the sample and the whole resonator: [28]

$$\eta = \frac{\int_{\text{sample}} B_1 \mathrm{d}V}{\int_{\text{resonator}} B_1 \mathrm{d}V}\qquad(2.71)$$

3 Scope of the Thesis

As discussed in chapter 2.3.3.5 the three-spin contribution depends on the angle between the two inter-spin vectors $\overrightarrow{KL}$ and $\overrightarrow{KM}$, while the two-spin contribution is independent from this angle. It should thus be possible to extract the angular information from the three-spin contribution. The goal of this thesis is to find an experimental method with which this is in principal possible. Therefore, various approaches were tested, which shall be briefly introduced in this section.

First, the dependency of the three-spin and the two-spin contribution on γ (see figure 2.15) was simulated. Two different cases were considered. For the first simulation, it was assumed that only the K spin can take the role of the observer spin. Hence, the change of the distance between the M and L spin by varying γ doesn't affect the two-spin and three-spin contribution, as ω_{ml} does not contribute. Consequently, the pure impact of changing the angle on the three-spin contribution could be examined. For the second simulation, every spin was allowed to be in the observer role. Thus, ω_{ml} contributes and the change of the distance between the M and L spin by varying γ affects the two-spin and three-spin contribution. The idea behind this experiment was to find out, how big the influence of the angle on the three-spin contribution is compared to its dependency on the distances. This simulation was done for the nitroxide triradicals.

Second, DEER experiments with chirps (about 250 MHz bandwidth) as pump pulses were carried out with the nitroxide triradicals. Narrow-band (about 15 MHz bandwidth) π pulses were used as observation pulses. The observation frequency was set on one edge of the nitroxide spectrum, while the rest was pumped with the chirp. Facing these conditions, the assumption that the inversion efficiency λ should be equal for every pumped spin seemed to be justified, as nearly the whole spectrum is inverted and the probability to find two spins in the small observation band was considered to be negligibly small. If this assumption was true, one can approximate that the pumping efficiency for all spins is independent from their orientation and therefore equal:

$$\lambda_l \approx \lambda_m \approx \lambda_k \tag{3.1}$$

With this approximation, the form factor (see equation 2.48), for the case that only the K spin can take the role of the observer spin, can be expressed as:

$$\begin{aligned}F_3(t) =& (1 - \lambda)^2 + \\ & (\lambda - \lambda^2)(\cos(\omega_{kl}t) + \cos(\omega_{km}t)) + \\ & \lambda^2 \cos(\omega_{km}t)\cos(\omega_{kl}t)\end{aligned} \tag{3.2}$$

and for the case that any of the three spins can be the observer spin:

$$\begin{aligned}F_3(t) =& (1 - \lambda)^2 + (\lambda - \lambda^2)\frac{1}{3}[\cos(\omega_{kl}t) + \cos(\omega_{km}t) + \cos(\omega_{lm}t)] \\ & + \lambda^2 \frac{1}{3}[\cos(\omega_{km}t)\cos(\omega_{kl}t) + \cos(\omega_{km}t)\cos(\omega_{lm}t) + \\ & \cos(\omega_{kl}t)\cos(\omega_{lm}t)].\end{aligned} \tag{3.3}$$

With these assumptions, one can define the two-spin contribution as:

$$P(t) = \frac{1}{3}[\cos(\omega_{kl}t) + \cos(\omega_{km}t) + \cos(\omega_{lm}t)] \tag{3.4}$$

and the three-spin contribution as:

$$T(t) = \frac{1}{3}[\cos(\omega_{km}t)\cos(\omega_{kl}t) + \cos(\omega_{km}t)\cos(\omega_{lm}t) + \cos(\omega_{kl}t)\cos(\omega_{lm}t)]. \tag{3.5}$$

Thus, equation 3.3 reduces to:

$$F(t) = (1 - \lambda)^2 + (\lambda - \lambda^2)P(t) + \lambda^2 T(t) \tag{3.6}$$

In the considered case, λ can be calculated from the experimental modulation depth (equation 2.51):

$$\Delta = 2\lambda - \lambda^2 \tag{3.7}$$

Knowing λ, one can separate $P(t)$ and $T(t)$, if at least two different DEER traces with different modulation depths are available. [1]

This experiment was performed to test the precision of the three-spin contributions extracted according to the made assumptions, as, in principle, only a precise extraction would allow for back-calculation of angle information from the experimental data. To evaluate the precision, simulations based on a fitted geometry model were used.

Third, DEER experiments with an adiabatic double chirp (two chirps each with about 250 MHz bandwidth) as pump pulse were carried out on the nitroxide radicals. The double chirp was designed out of two 48 ns, 250 MHz chirp pulses. One chirp was applied at lower frequencies than the observation frequency and one at higher frequencies. The spacing between the chirps and the observation frequency was 30 MHz. Under these experimental conditions, the requirements for neglecting orientation selection due to the pump spins should be fulfilled for observation frequencies everywhere in the nitroxide absorption spectrum. Thus, the influence of the orientation selection caused by the narrow observation band could be examined over the whole spectral range.

Besides the investigations in systems with three equal spin labels, DEER experiments with chirp pulses were performed on orthogonal spin labels. The model system was a Gd(III)-nitroxide triradical. As the Gd(III) and the nitroxides can be detected separately at Q band (see chapter 2.4.2), the whole nitroxide spectrum can be pumped and orientation selection of the pump pulse can be avoided completely. Moreover, Gd(III) is known to have a much less pronounced orientation selection due to the zero-field splitting [25]. Hence, this systems seems to be promising to exclude orientation selection.

4 Materials and Methods

4.1 Sample Preparation

All commercially available chemicals were used as received. Perdeuterated glycerol (98%, d8) was purchased from Isotec. Perdeuterated o-terphenyl was synthesized by Herbert Zimmermann, MPI for Medical Research, Heidelberg, Germany.

4.1.1 Nitroxide Radicals

The nitroxide radicals were diluted with perdeuterated o-terphenyl in such a manner that a spin concentration of $200 \cdot 10^{-6}$ mol/l was achieved. The mixtures were then pestled and filled into a EPR tubes. After this they were molten by heating to 70° C with a heat gun. Then they were immediately placed into the EPR probehead that was pre-cooled to 50 K to obtain glassy frozen solutions. The samples were stored at 277 K. The melting and shock-freezing procedure was repeated before every new experimental block.

4.1.2 Gd(III)-Nitroxide Radical

The investigated Gd(III)-nitroxide radical was dissolved in a 1:1 mixture of perdeuterated glycerol and D_2O with a nitroxide spin concentration of $150 \cdot 10^{-6}$ mol/l and a Gd(III) spin contration of $75 \cdot 10^{-6}$ mol/l. The sample preparation was done by Luca Garbuio and is described elsewhere [32]. The sample was stored at liquid nitrogen temperature.

4.2 Instrumentation

All experiments were performed on a Bruker ELEXSYS Q band pulse spectrometer. The pulses were generated with an Agilent M8190 12 GSa/s arbitrary waveform generator (AWG). The AWG provided two channels for the in-phase (I) and quadrature (Q) component. The AWG pulses were generated at a intermediate frequency (IF) of about 1.6 GHz. These pulses were then upconverted by a 8 GHz local oscillator (LO) (Nexyn NXPLOS). This upconversion scheme resulted in one sideband at around 9.6 GHz, which was then amplified (Meuro MBM08012G2423). The sideband was then fed into the incoherent ultra-wide band (UWB) channel of the spectrometer (see figure 4.1). Reflections were isolated by an X band isolator. The pulses were then translated to Q band and amplified by a high power traveling wave tube amplifier before they entered the resonator. Pulse generation was triggered by the Elexsys console. The AWG was controlled via a homemade MatLab-program running on an external computer.

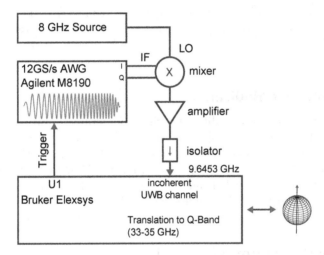

Figure 4.1: Schematic drawing of the connection of the AWG to the pulse EPR spectrometer.

As a resonator a homebuilt TE102 rectangular resonator suitable for oversized sample tubes up to 3.0 mm diameter was used. [33] The Q_L value was adjusted to the minimum possible value for this resonator (about 110) to achieve maximum resonator bandwidth.

4.3 Experiments

The experiments on the nitroxide radicals were carried out at 50 K, whereas the experiments on the Gd(III)-nitroxide radical were performed at 10 K. All experiments were done with sample tubes of 3.0 mm diameter.

4.3.1 Measuring of the Absorption Spectrum

We measured the absorption spectrum of the different spin labels with an ESE experiment (see chapter 2.3). We used a 32 ns π pulse and a 16 ns $\frac{\pi}{2}$ pulse in a field range of 200 to 400 G for the nitroxide spin labels. For the Gd spin label a 24 ns π pulse and a 12 ns $\frac{\pi}{2}$ pulse were used in a field range of 4000 G. τ was set to 400 ns in both cases.

4.3.2 DEER Experiments

4.3.2.1 Nitroxide Radicals

128 ns π pulses and a 64 ns $\frac{\pi}{2}$ pulse were used for the DEER experiments on the nitroxide radicals on the observer frequency. τ_1 was set to 400 ns and τ_2 to 6000 ns.

For the DEER experiment with a chirp pump pulse one 48 ns, 250 MHz chirp pulse was used as pump pulse. On the reference compound (MSA347 diradical), additional DEER experiments with a 96 ns, 250 MHz chirp pump pulse were performed. Moreover, monochromatic pump pulses, with flip angles either calibrated to π or reduced intentionally, were used under the same conditions as the chirps as references. For the DEER experiment with a double chirp as pump pulse two 48 ns, 250 MHz chirp pulses were used as pump pulses. The pulses were applied right after each other, so that the total pulse length was 96 ns. With the double chirp a total bandwidth of 500 MHz could be achieved.

4.3.2.2 Gd(III)-Nitroxide Radical

For the Gd(III)-nitroxide DEER, the Gd(III) spin system was observed while the nitroxide labels were pumped. 12 ns π pulses and 12 ns $\frac{\pi}{2}$ pulses were used on the observer frequency and 48 ns chirp pulses with a bandwidth of 250 MHz were used as pump pulses. Besides, a 17 ns monochromatic pump pulse, with flip angle calibrated to π, was used under the same conditions. τ_1 was set to 400 ns and τ_2 to 6000 ns.

4.3.3 Echo Reduction Measurements

To measure the strength of the echo reduction effect for different pulse positions, a fixed observation frequency was chosen and a DEER experiment was setup. In contrast to a real DEER experiment the position of the pump pulse was not varied in time, but fixed such that the maximum of the DEER echo was detected. This experiment was repeated for different start and end frequencies of the chirp pulse. A justifiable position of the chirp pulse and of the observation frequency with not too high echo reduction, not too low observation intensity and an optimal pumping efficiency could be found with this procedure.

4.4 Determination of the Resonator Profile

As mentioned above (see chapter 2.1.4), the nutation frequency of an on-resonance pulse is directly proportional to the amplitude of its magnetic field component B_1 (see equation 2.7). Thus, one can evaluate how good a specific frequency is coupled into the resonator, if one knows the nutation frequency of the according on-resonance pulse. In chapter 2.3.2, an experiment for determining the nutation frequency is explained. This experiment was repeated for different frequencies. The magnetic field was adjusted before every measurement to ensure that the pulses acts always on the absorption maximum of the nitroxide spectrum. In this way, the resonance profile of the resonator could be determined. This experiment was mostly performed in the frequency range from 33.500 to 34.900 GHz, with an increment of 25 MHz, τ was set to 400 ns.

4.5 Pulse Characterization

4.5.1 Setting of the Pulse Amplitude and Length

The length of the used monochromatic pump pulses was calibrated using the same experiment as for the determination of the resonator profile (see chapter 4.4). The length of the chirp pump pulse was chosen based on a series of test DEER measurements on the Gd-nitroxide radical (see figure 4.2). In this test series, 250 MHz chirps with different length were used as pump pulses on the nitroxides. As a result, it was found that the modulation depth didn't improve anymore, if the chirp length is raised further from 48 ns. Therefore, a length of 48 ns can be considered as sufficient to efficiently pump nitroxide spins with our instrumentation.

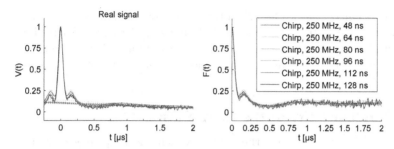

Figure 4.2: Impact of the chirp pump pulse length on the modulation depth. Shown is the outcome of experimental DEER measurements on the Gd-nitroxide radical. Left figure: Real part of the phase corrected DEER signals (doted lines: fitted background). Right figure: Form factors (= phase and background corrected experimental DEER signals).

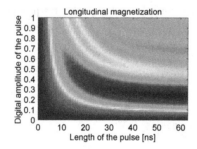

Figure 4.3: Pseudocolor plot of the longitudinal magnetization of a nitroxide test sample in dependency of the digital amplitude and length of a pulse. The intensity values are displayed in a color code. Color code for the print version: The darker the color, the higher is the absolute value of the magnetization. Color code for the e-book: Dark red is used for the highest and dark blue for the lowest magnetization values.

As it is essential to generate DEER data with different λ values to extract the two-spin and three-spin contributions, the amplitude of the pump pulse have to be chosen such that the modulation depth of the corresponding DEER signals varies clearly. Therefore, a two-dimensional nutation experiment was performed on a nitroxide test sample (see figure 4.3). In this experiment the dependency of the flip angle of a pulse on its amplitude and on its length was investigated. As a conclusion of this, the typical amplitudes of the used monochromatic pulses were scaled with 1.0, and 0.28. The amplitudes of the used chirp pulses were adjusted with DEER test measurements. The optimum chirp amplitudes for nitroxide-nitroxide DEER were found to be scaled by 1.0, 0.38, 0.28 and 0.127. For Gd-nitroxide DEER the best amplitudes were found to be 1.0, 0.11 and 0.01. In addition to the reduction of the digital scale, the analog amplitude of the last chirp was attenuated by 10 dB[1].

4.5.2 Adiabaticity of the Chirp Pulses

As stated in chapter 2.1.5 the rate at which the frequency of a pulse is swept must be small compared to the m.w. amplitude to fulfill the adiabatic condition. In order to quantify this criterion the adiabaticity factor Q is defined as:

$$Q = \frac{\omega_1^2 t_p}{2\pi\Delta f} \tag{4.1}$$

whereat t_p is the pulse length and Δf the frequency range of the pulse. [14, 34]

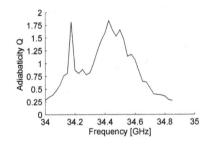

Figure 4.4: Adiabaticity Q of a 48 ns and 250 MHz chirp pulse as a function of its frequency: The spectrum was calculated by using the nutation frequencies per pulse frequency (see chapter 4.4) as ω_1 in equation 4.1.

Figure 4.4 shows the adiabaticity Q of a 48 ns and 250 MHz chirp pulse on the used spectrometer and resonator. The measurement was performed on a test sample. It is well known that Q should be above 5 to fulfill the adiabatic condition. [14] Therefore the used chirp pulses do not strictly fulfill the adiabatic condition. However, due to technical limitations, it is not possible to use chirp pulses with higher Q values for DEER at present.

---

[1]decibel

4.6 Processing of Experimental Data

4.6.1 Background Correction

As mentioned before, the signal measured in DEER experiments is the product of a form factor and a background factor (compare equation 2.47). All presented form factors were extracted from the measured data by using the program DeerAnalysis [35], version 2013.2 [36]. The background correction was performed by fitting the background factor with formula 2.44. The dimensionality and the decay rate were set as fit parameters.

4.6.2 Phase Correction

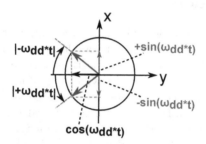

Figure 4.5: DEER in the high temperature limit: Illustration of the positions of the magnetization vectors in the rotating frame when the signal is detected.

Our experimental conditions fulfill the requirements of the high temperature limit (magnetic field strength below 6 T and temperature above 5 K). Therefore, assuming an infinitely broad distribution of resonance offsets, we can anticipate that we have almost equal numbers of spins oriented parallel and antiparallel to the magnetic field. As a consequence after the pump pulse one half of the observed spins will rotate at the frequency $-\omega_{dd}$ in the rotating frame whereas the other half will rotate at a frequency of $+\omega_{dd}$ (as shown in figure 4.5). The sum of the magnetizations of these two sub-ensembles is the observed echo signal:

$$s(t) = \frac{1}{2}[exp(i\omega_{dd}t) + exp(-i\omega_{dd}t)] = cos(\omega_{dd}t) \qquad (4.2)$$

This result means that one channel of the quadrature-detected transverse magnetization contains the full signal, while the other one is zero. [37]

However, exact adjustment of the phase of the echo signal can be difficult if the signal is weak and it is know that DEER pump pulses can cause dynamic phase shifts [38]. Moreover, the phase may drift during long accumulation times. [35] It is therefore necessary to record both signals and perform phase correction after the measurement. The presented data was phase corrected with the automatic phase correction in the DeerAnalysis program. The optimum phase is determined by minimizing the root mean square deviation from zero of the imaginary signal for the last three quarters of the data. [35] This is done before the background correction.

4.7 Simulation of the Three-Spin and Two-Spin Contribution

For the simulation of the two-spin and three-spin contribution, the input parameters were two side lengths, the angle between those sides and the standard deviation. In a first step, the third side length was calculated. Then the real simulation started, for which a uniform standard deviation for all side lengths was assumed and incorporated as a Monte-Carlo loop. The free-electron Zeeman factor, g_e, was used to calculate ω_{dd} for the pump and observer spins. Orientation selection was not taken into account. The simulation was performed for the time-domain signal.

5 Results and Discussion

In 2009 *Jeschke, G. et al.* proved that it is, in principle, possible to extract inter-spin angles of small triradicals with not too complex structures by fitting a geometry model to the experimental two-spin and three-spin contribution [1]. For more complex molecules, e.g. proteins, however it might not be possible to fit the experimental data with a simple geometry model. Therefore, a way of extracting the angular information from the higher-order terms of the experimental DEER signal has to be found. The theoretical dependency of those higher-order terms on the inter-spin angle is known and has been discussed for the second-order term (three-spin contribution) in chapter 2.3.3.5. However, to extract the angle information, the terms of the experimental DEER signal have to be separated from each other. In the following, different experimental approaches and procedure will be tested on systems with three spins for their applicability to generate appropriate DEER data, from which the two-spin and three-spin contribution can be extracted by using the approximations made in chapter 3.

5.1 Nitroxide-Nitroxide DEER

5.1.1 Simulation of the Angle Dependency of T and P

In this section the results of the simulation of the two-spin and three-spin contribution for the investigated nitroxide triradicals are presented. To find out how big the influence of the inter spin angles is in comparison to the distances, two different cases were simulated. First, the two-spin and three-spin contribution were calculated for different angles with the restriction that only the spin in between the two inter-spin vectors that enclose the angle is observed and the other two spins are pumped. If the nomenclature of figure 5.1 is used, this corresponds to the case that only the K spin is observed, while the M and L spin are pumped. In this case the change of the third distance $(\overrightarrow{ML})$ doesn't affect the two-spin and three-spin contribution. Thus the pure influence of the inter-spin angle can be investigated. For the second simulation every spin (K, M and L) was set into the observer and pump positions. Thus the change of the distance $\overrightarrow{ML}$ affects the two-spin and three-spin contribution.

The comparison of both simulation provides an idea of how big the influence of the angles is in comparison to the distances.

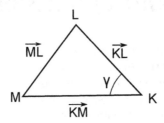

Figure 5.1: Geometry of a three-spin system to visualize the used nomenclature. In the first simulation only the K spin is observed while M and L are pumped only. In the second simulation every spin can take every roll equally. γ is the angle that is changed during the simulations.

5.1.1.1 Simulation with a Fix Observer Spin

For the simulation, geometry information about the investigated nitroxide triradicals gained by DEER experiments and basic chemical knowledge were used. The angle was just varied in a range that seemed to be realistic by looking at the structure.

Triradical T36d   In figure 5.2 the time-domain signal of the simulated two-spin and three-spin contribution of T36d are presented. The angle was varied in the range from 55° to 65° as the molecule structure suggests an angle of 60°. As predicted, the two-spin contribution does not change with the angle. In the three-spin contribution, small but clear changes can be seen.

Figure 5.3 shows pseudocolor plots of the dipolar spectra of the two-spin and three-spin contribution versus the angle. In the frequency domain, one clear trend can be figured out for the three-spin contribution. The percentage of small frequencies is increasing with an increasing angle, as the spectra are getting sharper in the small frequency region ($|\omega/2\pi| < 1$). This means that the red area around zero is getting narrower and darker in the pseudocolor plot.

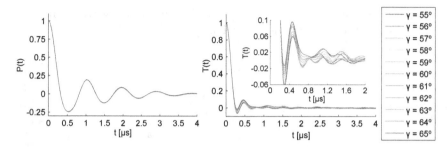

Figure 5.2: Simulated time-domain signal of the two-spin and three-spin contribution of T36d with fixed observer spin: Both inter-spin vectors were set to 3.64 nm with a standard deviation of 0.1 nm.

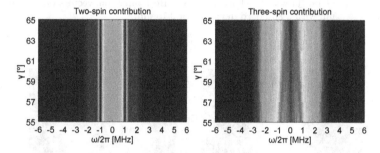

Figure 5.3: Pseudocolor plot of the dipolar spectra of the simulated two-spin and three-spin contribution of T36d versus the inter-spin angle with fixed observer spin: Both inter-spin vectors were set to 3.64 nm with a standard deviation of 0.1 nm. The intensity values are displayed in a color code. Color code for the print version: Black is used for the highest and the lowest intensity values. Color code for the e-book: Dark red is used for the highest and dark blue for the lowest intensity values. For the individual dipolar spectra per inter-spin angle, see figure 1 in the appendix on page 93.

Triradical TR011 In figure 5.4 the time-domain signal of the simulated two-spin and three-spin contribution of TR011 are presented. The angle was varied in the range from 55° to 75° as the molecule structure suggests an angle of around 70°. The two-spin contribution is again found to be independent from the angle. In the three-spin contribution changes that are analogue to the investigations on T36d are observed.

Figure 5.5 shows pseudocolor plots of the dipolar spectra of the two-spin and three-spin contribution versus the angle. In the frequency domain of the three-spin contribution of TR011 the percentage of small frequencies is also increasing with an increasing angle, as the spectra are getting narrower and darker in the small frequency region ($|\omega/2\pi| < 1$).

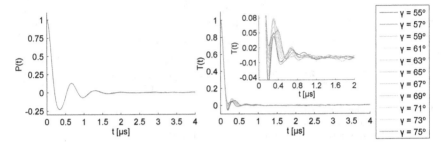

Figure 5.4: Simulated time-domain signal of the two-spin and three-spin contribution of TR011 with fixed observer spin: Both inter-spin vectors were set to 3.15 nm with a standard deviation of 0.15 nm.

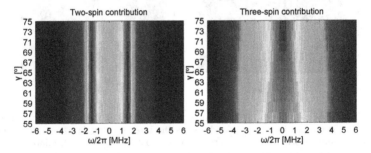

Figure 5.5: Pseudocolor plot of the dipolar spectra of the simulated two-spin and three-spin contribution of TR011 versus the inter-spin angle with fixed observer spin: Both inter-spin vectors were set to 3.15 nm with a standard deviation of 0.15 nm. The intensity values are displayed in a color code. Color code for the print version: Black is used for the highest and the lowest intensity values. Color code for the e-book: Dark red is used for the highest and dark blue for the lowest intensity values. For the individual dipolar spectra per inter-spin angle, see figure 2 in the appendix on page 93.

5.1.1.2 Simulation with a Variable Observer Spin

For the simulations with variable observer spins the same distances, standard deviations and angles were assumed as for the simulation with fix observer spin.

Triradical T36d As the third distance $(\overrightarrow{ML})$ contributes now to the simulated DEER signal, the two-spin contribution also changes with the angle. Moreover, the three-spin contribution is influenced by the third distance. Thus, the signals of the simulated two-spin and three-spin contribution (figure 5.7) of T36d look different than those presented in the chapter before. However, the trend that the red area around zero is getting darker in the pseudocolor plot (see figure 5.6) in the small frequency region ($|\omega/2\pi| < 1$) can still be observed.

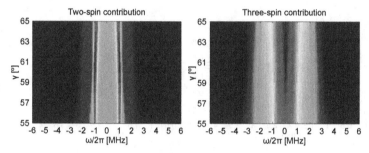

Figure 5.6: Pseudocolor plot of the dipolar spectra of the simulated two-spin and three-spin contribution of T36d versus the inter-spin angle with variable observer spin: Both inter-spin vectors were set to 3.64 nm with a standard deviation of 0.1 nm. The intensity values are displayed in a color code. Color code for the print version: Black is used for the highest and the lowest intensity values. Color code for the e-book: Dark red is used for the highest and dark blue for the lowest intensity values. For the individual dipolar spectra per inter-spin angle, see figure 3 in the appendix on page 94.

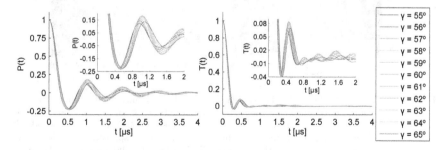

Figure 5.7: Simulated time-domain signal of the two-spin and three-spin contribution of T36d with variable observer spin: Both inter-spin vectors were set to 3.64 nm with a standard deviation of 0.1 nm.

Triradical TR011   The two-spin contribution of TR011 now also changes
with the angle (see figure 5.8). As the TR011 triradical is not symmetric like
the T36d radical, it is no surprise that the signals of the simulated two-spin
and three-spin contribution look pretty different than the ones in the section
before. This is a must, as the spins differ in their surrounding. However, the
trend that the amount of small frequencies is increasing with an increasing
angle can now be observed for the two-spin and the three-spin contribution
(see figure 5.9).

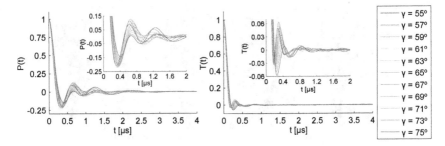

Figure 5.8: Simulated time-domain signal of the two-spin and three-spin contribution
of TR011 with variable observer spin: Both inter-spin vectors were set to
3.15 nm with a standard deviation of 0.15 nm.

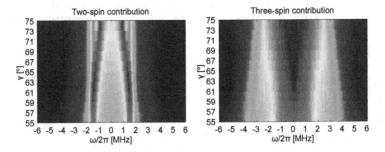

Figure 5.9: Pseudocolor plot of the dipolar spectra of the simulated two-spin and three-
spin contribution of TR011 versus the inter-spin angle with variable observer spin: Both inter-spin vectors were set to 3.15 nm with a standard
deviation of 0.15 nm. The intensity values are displayed in a color code.
Color code for the print version: Black is used for the highest and the lowest intensity values. Color code for the e-book: Dark red is used for the
highest and dark blue for the lowest intensity values. For the individual
dipolar spectra per inter-spin angle see figure 4 in the appendix on page
94.

5.1.1.3 Conclusion

The result of this section is that the changes in the three-spin contribution that are directly related to variations in the angle have about the same magnitude as those changes that are caused by variations in the third side length. Consequently it should be pretty difficult to extract the angle information from the three-spin contribution, especially for unsymmetrical radicals. To solve this task, very well resolved experimental data is needed. On the one side a high signal-to-noise ratio is necessary, this means high magnetic fields have to be used to split the electronic levels nicely. On the other side experimental data with high λ values is essential, as the contribution of T to the measured signal scales with λ^2. However the higher the magnetic field, the broader the spectrum and thus the more difficult it is to excite all pumped spins.

5.1.2 DEER with a Chirp Pump Pulse

To achieve a sufficient signal-to-noise ratio, the following measurements were performed at Q band (about 35 GHz and 12500 G). For high λ values and to be able to exclude orientation selection (compare chapter 3), it is important to flip all pumped spins homogeneously over the whole spectrum. Therefore chirps with a sufficiently high bandwidth of 250 MHz were employed as pump pulses.

5.1.2.1 Triradical T36d

To flip most of the pumped spins, the observation frequency had to be set on one edge of the NO spectrum, while the rest was pumped with the chirp. To fulfill this criteria the observation frequency can be set either to the upper or to the lower edge of the nitroxide spectrum. DEER experiments for both possible experimental setups were performed (see figure 5.10). With these conditions, one should be able to ignore orientation selection at the pumped spins (compare chapter 3). With every experimental setup, six different DEER traces were measured. In four measurements, chirp pulses with different digital amplitudes were used in order to generate experimental data with different modulation depth values.

In the other two measurements monochromatic pump pulses, with flip angles either calibrated to π (scale $= 1.0$) or reduced intentionally (scale $= 0.28$), were used to have some reference DEER trace for evaluating the quality of the DEER signals achieved with chirp pump pulses. All experimental details are summarized in table 5.1. The experimental DEER data on T36d is shown in the figures 5.11 (experimental setup I) and 5.12 (experimental setup II).

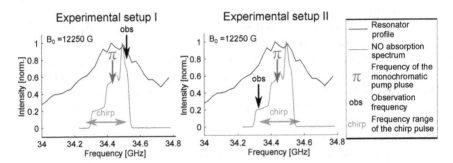

Figure 5.10: Experimental setup I and II of the DEER experiments on T36d. Shown is the position of the nitroxide absorption spectrum relative to the resonance profile of the resonator at a given external static magnetic field B_0 and how the pulse positions were set up.

Table 5.1: Experimental setup I and II of the DEER experiments on T36d.

parameter	experimental setup I	experimental setup II
external magnetic field	12250 G	12250 G
observation frequency	34.51 GHz	34.33 GHz
frequency range of the chirp pump pulse	34.26-34.51 GHz	34.33-34.58 GHz
length of the chirp pump pulse	48 ns	48 ns
frequency of the monochromatic pump pulse	34.43 GHz	34.43 GHz
length of the monochromatic pump pulse	12 ns	12 ns

Graph **A** shows the real part of the phase-corrected DEER signal of the measurements on T36d with experimental setup I. As the chirp pulse is about four times longer than a conventional monochromatic pump pulse, the zero point of the DEER trace might be hard to identify by the DeerAnalysis program. By comparing the traces of the monochromatic pulses with those

of the chirps, such problems can be excluded here. The imaginary part of
the phase-corrected DEER signal is shown in graph **B**.

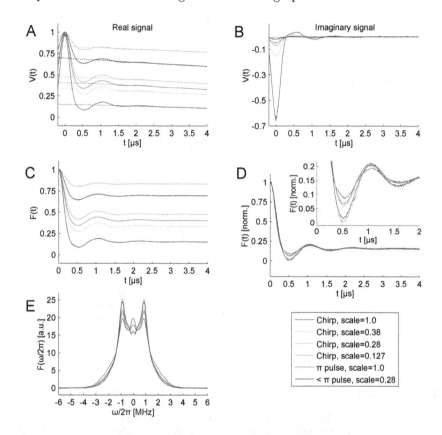

Figure 5.11: Experimental DEER data for T36d with pulse and B-field setting I (for
details see table 5.1). **A)** Real part of the phase-corrected DEER signals (dotted lines: fitted background). **B)** Imaginary part of the phase-
corrected DEER signals. **C)** Form factors (= phase- and background-
corrected experimental DEER signals, dotted lines: fit from which the
dipolar spectra are calculated). **D)** Normalized form factors (= phase-
and background-corrected experimental DEER signals that are scaled
such that they have the same modulation depth). **E)** Dipolar spectra (=
Fourier transformation of the corresponding form factor).

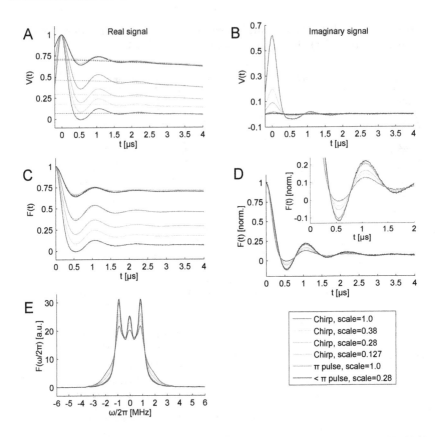

Figure 5.12: Experimental DEER data for T36d with pulse and B-field setting II (for details see table 5.1). **A)** Real part of the phase-corrected DEER signals (dotted lines: fitted background). **B)** Imaginary part of the phase-corrected DEER signals. **C)** Form factors (= phase- and background-corrected experimental DEER signals, dotted lines: fit from which the dipolar spectra are calculated). **D)** Normalized form factors (= phase- and background-corrected experimental DEER signals that are scaled such that they have the same modulation depth). **E)** Dipolar spectra (= Fourier transformation of the corresponding form factor).

As explained above (compare chapter 4.6.2) the imaginary signal should be zero after the phase correction. This is definitely not the case here. The imaginary signal scales with the modulation depth. Moreover, it is remarkable that even the DEER traces measured with monochromatic pump pulses show a nonzero imaginary signal. By comparing the two different experimental setups, it is striking that the imaginary signal is positive for experimental setup I and negative for setup II. This leads to the conclusion that it might depend on the relative position of the pump pulse with respect to the observation frequency. However, at this point it is not possible to interpret these findings. Graph **C** shows the form factors which are used to extract the two-spin and three-spin contribution. In graph **D** they are normalized such that they superimpose. This is done in order to be able to compare them easily. The higher the modulation depth the faster is the decay and the smaller is the oscillating amplitude. This corresponds to more higher frequency content in the form factors with high modulation depth.

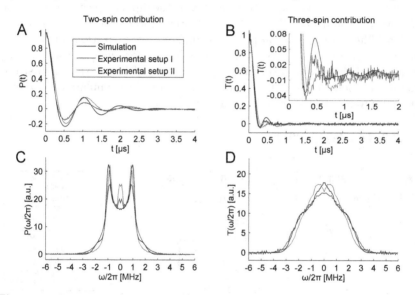

Figure 5.13: Experimental and simulated two-spin and three-spin contributions of T36d. Both inter-spin vectors were set to 3.64 nm, γ was set to $60°$ and a standard deviation of 0.15 nm was assumed. Only the DEER traces of the experiments in which a chirp pump pulse was used were include in the calculation of the experimental two-spin and three-spin contribution. **A)** Time-domain signal of the two-spin contribution. **B)** Time-domain signal of the three-spin contribution. **C)** Dipolar spectra of the two-spin contribution. **D)** Dipolar spectra of the three-spin contribution.

Graph **E** shows the dipolar spectra of the measured DEER traces. Those confirm the assumption that the contribution of higher frequencies increases with the modulation depth. These higher frequency contributions can be partially assigned to the three-spin contribution, as T contains sum frequencies. Thus, this proves that the higher the modulation depth, the more three-spin contribution is contained in the DEER signal.

Figure 5.13 shows the experimental and simulated two-spin and three-spin contribution of T36d. The experimental T and P data has been calculated from the DEER experiments presented above according to equation 3.6 on page 40. Only those DEER traces that were generated by using a chirp pump pulse were included in the calculation. The time-domain signals both of the two-spin (graph **A**) and the three-spin contribution (graph **B**) extracted from the DEER trace of setup I decay faster than those of setup II. This means that the contribution of high frequencies is larger in the DEER traces of setup I. This observation is also verified by the dipolar spectra of P (graph **C**) and T (graph **D**). This is a hint at orientation selection, as it means that at different observation positions, different configurations of the same molecule are measured. As it can be expected for such a case [21], the simulated P looks somehow like a combination of both experiments, but doesn't fit good to any of the experimental data. The simulated T however fits quite well with the experimental one of setup I.

5.1.2.2 Triradical TR011

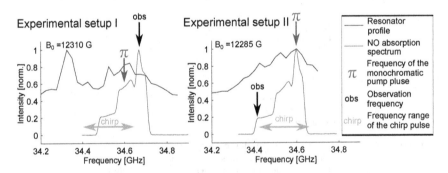

Figure 5.14: Experimental setup I and II of the DEER experiments on T36d. Shown is the position of the nitroxide absorption spectrum relative to the resonance profile of the resonator at a given external static magnetic field B_0 and how the pulse positions were set up.

The experimental setups of the DEER experiments on triradical TR011 were analogous to the ones on T36d. We performed twelve DEER experiments on TR011 in two different experimental setups, six per setup. For experimental setup I, the observation frequency was set to the higher edge of the nitroxide spectrum while the rest was pumped. For setup II, the pulse setting was exactly the other way round (see figure 5.14). Per setup, four DEER experiments were performed using a chirp as pump pulse and two with a monochromatic pump pulses with a flip angle either calibrated to π (scale = 1.0) or reduced intentionally (scale = 0.28). The digital amplitudes of the pump pulses were analogously scaled to those of the DEER experiments on T36d. For experimental details see table 5.2.

Table 5.2: Experimental setup I and II of the DEER experiments on TR011.

parameter	experimental setup I	experimental setup II
external magnetic field	12310 G	12285 G
observation frequency	34.68 GHz	34.42 GHz
frequency range of the chirp pump pulse	34.40-34.65 GHz	34.43-34.68 GHz
length of the chirp pump pulse	48 ns	48 ns
frequency of the monochromatic pump pulse	34.60 GHz	34.60 GHz
length of the monochromatic pump pulse	16 ns	13 ns

The experimental DEER traces that were measured with setup I and II are presented in figure 5.15 and 5.16. Graph **A** shows the real part of the phase-corrected DEER signals. First of all, it is important to mention that all DEER traces have a well defined zero point. The striking observation by comparing the **A** graphs of figure 5.15 and 5.16 is that the ones of setup I have a clearly smaller modulation depth than those of setup II. This might be caused by the pulse positions, as in setup I less of the spectrum is pumped by the chirp pulses than in setup II (compare figure 5.14). Another reason might be the coupling of the resonator, which was lower for setup I, as can be seen by the length of the monochromatic π pulses in table 5.2. The imaginary part of the phase-corrected DEER signal (graph **B**) is again nonzero.

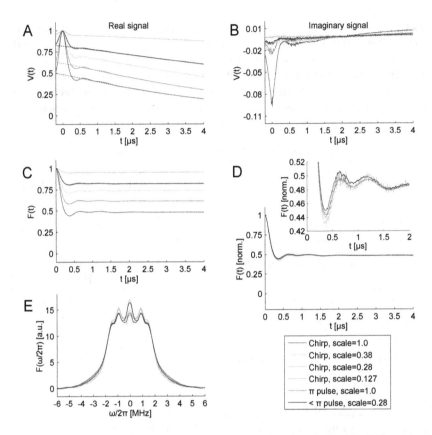

Figure 5.15: Experimental DEER data for TR011 with pulse and B-field setting I
(for details see table 5.1). **A)** Real part of the phase-corrected DEER
signals (dotted lines: fitted background). **B)** Imaginary part of the phase-
corrected DEER signals. **C)** Form factors ((= phase- and background-
corrected experimental DEER signals, dotted lines: fit from which the
dipolar spectra is calculated). **D)** Normalized form factors (= phase-
and background-corrected experimental DEER signals that are scaled
such that they have the same modulation depth). **E)** Dipolar spectra (=
Fourier transformation of the corresponding form factor).

It is negative for setup I and positive for setup II. However the maximal
amplitudes of the imaginary signal in setup I are considerably smaller than
in setup II. This leads to the conclusion that the amount of imaginary signal
is related to the modulation depth. The forms factors are presented in graph
C and **D**. The scaled form factors of experimental setup I are very similar.

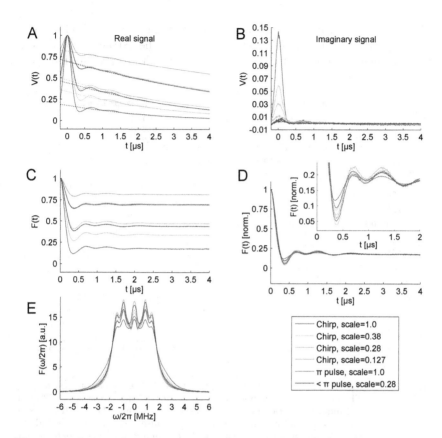

Figure 5.16: Experimental DEER data for TR011 with pulse and B-field setting II
(for details see table 5.1). **A)** Real part of the phase-corrected DEER
signals (dotted lines: fitted background). **B)** Imaginary part of the phase-
corrected DEER signals. **C)** Form factors (= phase- and background-
corrected experimental DEER signals, dotted lines: fit from which the
dipolar spectra is calculated). **D)** Normalized form factors (= phase-
and background-corrected experimental DEER signals that are scaled
such that they have the same modulation depth). **E)** Dipolar spectra (=
Fourier transformation of the corresponding form factor).

This means that the percentage of three-spin contribution is very similar
in the individual DEER traces. The form factors of experimental setup II
show a different decay rate. This corresponds to different percentages of
three-spin contribution in the single traces. The dipolar spectra (graph **E**)
confirm this conclusions. The ones of setup I show only minor differences,

while the ones of setup II feature a clear shift to higher frequencies for traces
with larger modulation depths. This means that the extraction of T from
the experimental data of setup I is more difficult and less precise than from
setup II.

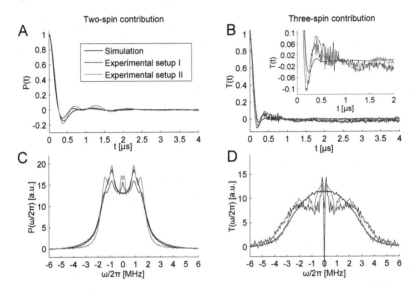

Figure 5.17: Experimental and simulated two-spin and three-spin contributions of
TR011. Both inter-spin vectors were set to 3.15 nm, γ was set to 72° and
a standard deviation of 0.25 nm was assumed. Only the DEER traces of
the experiments in which a chirp pump pulse was used were include in
the calculation of the experimental two-spin and three-spin contribution.
A) Time-domain signal of the two-spin contribution. **B)** Time-domain
signal of the three-spin contribution. **C)** Dipolar spectra of the two-spin
contribution. **D)** Dipolar spectra of the three-spin contribution.

Figure 5.17 shows the two-spin and three-spin contribution of TR011. The
experimental T and P data has been calculated analogously to T36d only
from those DEER traces that were generated by using a chirp pump pulse.
As expected, the resolution of the three-spin contribution (graph **B** and **D**)
extracted from setup I is quite low compared to the other extracted signals.
However it is sufficiently good to draw qualitative conclusions. The time-
domain signals of the two-spin (graph **A**) and the three-spin contribution
(graph **B**) of setup I decay faster than those of setup II. This corresponds
to a shift of the dipolar spectra of setup I to higher frequencies compared
to the ones of setup II (for $P(\omega/2\pi)$ (graph **C**) and $T(\omega/2\pi)$ (graph **D**)).

This leads again to the conclusion that at different observation positions, different configurations of the same molecule are measured. Consequently, the simulated P and T don't fit too well to any of the experimental data.

5.1.2.3 Diradical MSA347

To figure out where the non-zero imaginary signal has its origin and to exclude that it is a three-spin effect, test measurements on a diradical were performed. In the first experiment series, it was examined how the imaginary signal is influenced by the chirp amplitude. For this, the observation frequency was set to the lower edge of the nitroxide spectrum, while the rest of the spectrum was pumped with a 250 MHz, 48 ns chirp pulse. In the second experiment series, the relation between the amplitude of the imaginary signal and the pulse length was investigated. For this, the same setup was used as for the first series of experiments (see figure 5.18), the pump pulse length was doubled to 96 ns and the bandwidth was kept constant at 250 MHz.

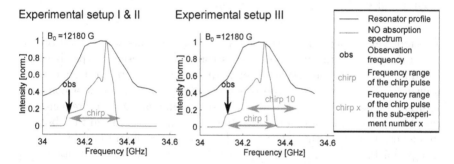

Figure 5.18: Experimental setup I, II and III of the DEER experiments on MSA347. Shown is the position of the nitroxide absorption spectrum relative to the resonance profile of the resonator at a given external static magnetic field B_0 and how the pulse positions were set up.

Table 5.3: Experimental setup I, II and III of the DEER experiments on MSA347.

parameter	experimental setup I	experimental setup II	experimental setup III
external magnetic field	12180 G	12180 G	12180 G
observation frequency	34.13 GHz	34.13 GHz	34.13 GHz
frequency range of the chirp pump pulse	34.13-34.38 GHz	34.13-34.38 GHz	10 different chirp pulses in the range of 34.13-34.38 GHz to 34.22-34.47 GHz were used for the experiment
length of the chirp pump pulse	48 ns	96 ns	48 ns

The motivation behind the third series of experiments was to find out, if the non-zero imaginary signal is so strong for the used chirps, because the difference between the observer frequency and the pump frequency is unusually small. For this, the observer frequency was set at a constant position at the low edge of the nitroxide spectrum and the start frequency of the chirp pump pulse was shifted in steps of 10 MHz away from the observer frequency. Details of the experimental setups can be found in table 5.3.

In figure 5.19 the results of the DEER experiments with experimental setup I on MSA347 are presented. Graph **B** shows the imaginary part of the phase-corrected DEER signal. It clearly scales with the pulse amplitude. The form factors (graph **C** and **D**) of all DEER traces show the same decay behavior. Thus, the dipolar spectra (graph **E**) look all identical. This is an expected result, as the signal of the diradical consists only of one dipolar frequency. Different λ values do not give rise to a different weighting of frequency components.

The results of the DEER experiments with experimental setup II on MSA347 are shown in figure 5.20. The real part of the phase-corrected DEER signal is shown in graph **A**. The pulses with a scale of 1.0 to 0.6 have all a modulation depth of about 81-82 %. Their imaginary signals, however, look pretty different.

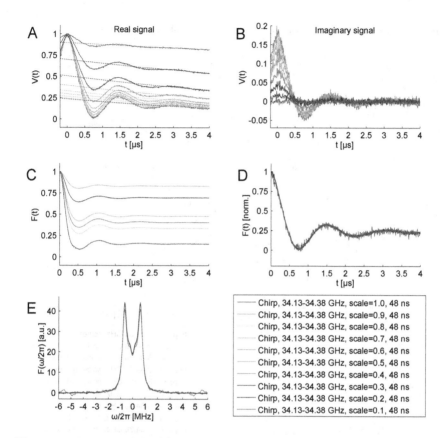

Figure 5.19: Experimental DEER data for MSA347 with pulse and B-field setting I (for details see table 5.1). **A)** Real part of the phase-corrected DEER signals (dotted lines: fitted background). **B)** Imaginary part of the phase-corrected DEER signals. **C)** Form factors (= phase- and background-corrected experimental DEER signals, dotted lines: fit from which the dipolar spectra is calculated). **D)** Normalized form factors (= phase- and background-corrected experimental DEER signals that are scaled such that they have the same modulation depth). **E)** Dipolar spectra (= Fourier transformation of the corresponding form factor).

The maximal amplitude of the imaginary signal increases with the pulse amplitude from scale 0.1 to scale 0.4. For pulses with higher scales the imaginary signal decreases and turns even negative. Besides this, the form factors (graph **C** and **D**) and the dipolar spectra (graph **E**) look exactly equal. Thus, the imaginary signal is related to long pulse lengths and does not appear to influence the outcome of the DEER experiments.

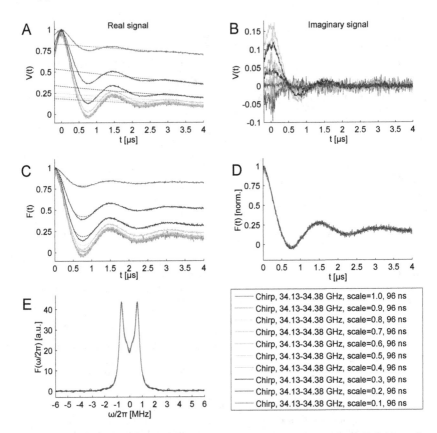

Figure 5.20: Experimental DEER data for MSA347 with pulse and B-field setting
II (for details see table 5.1). **A)** Real part of the phase-corrected DEER
signals (dotted lines: fitted background). **B)** Imaginary part of the phase-
corrected DEER signals. **C)** Form factors (= phase- and background-
corrected experimental DEER signals, dotted lines: fit from which the
dipolar spectra is calculated). **D)** Normalized form factors (= phase-
and background-corrected experimental DEER signals that are scaled
such that they have the same modulation depth). **E)** Dipolar spectra (=
Fourier transformation of the corresponding form factor).

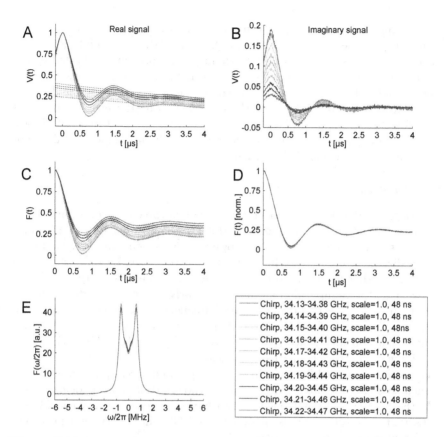

Figure 5.21: Experimental DEER data for MSA347 with pulse and B-field setting
III (for details see table 5.1). **A)** Real part of the phase-corrected DEER
signals (dotted lines: fitted background). **B)** Imaginary part of the phase-
corrected DEER signals. **C)** Form factors (= phase- and background-
corrected experimental DEER signals, dotted lines: fit from which the
dipolar spectra is calculated). **D)** Normalized form factors (= phase-
and background-corrected experimental DEER signals that are scaled
such that they have the same modulation depth). **E)** Dipolar spectra (=
Fourier transformation of the corresponding form factor).

The outcome of the measurements with experimental setup III on diradical
MSA347 is shown in figure 5.21. In this series of experiments the correlation
between the position of the pump pulse relative to the observation frequency
and the amount of imaginary signal was investigated. The modulation depth
and the amplitude of the imaginary signal decrease both constantly with an

increasing frequency gap between the observation frequency and the start frequency of the chirp pulse (see graph **A** and **B**). The modulation depth decreases because a smaller part of the nitroxide spectrum is affected by the pump pulse (compare experimental setup III in figure 5.18). Thus it cannot be excluded that the decrease in imaginary signal is caused by the decrease in modulation depth instead of the pulse position. However the decay of the form factors is again identical for every measurement and the dipolar spectra look the same. Thus the differences in the imaginary signal do not influence the outcome of the DEER experiment.

5.1.2.4 Conclusion

It was found that the experimental two-spin and three-spin contribution vary for different observation frequencies. Moreover non-zero phase-corrected imaginary signals could be observed. However, due to some test measurements on a diradical, one can exclude that those are caused by some three-spin effect. Further, it seems to be very likely that the imaginary signals are caused by the pump pulse, as they are clearly related to its duration and position with respect to the observation frequency. As the measurements on the diradical showed the same results independent of the imaginary signal, the conclusion was drawn that they are most probably artifacts that have no influence on the real part of the measured signal.

The reason for the differences in the two-spin and three-spin contributions, when changing the observation position, though could not be identified yet. As the weighting of the frequency contributions is found to vary for different observation positions, it is likely that at different observation positions, different conformations are observed.

5.1.3 DEER with a Double Chirp Pump Pulse

To examine orientation selection effect in the investigated nitroxide triradicals further, a new kind of pump pulse, a double chirp pulse, was developed. This double chirp consists of two separate 48 ns chirps with a bandwidth of 250 MHz each. The chirps are generated after each other so that the double chirp has a total length of 96 ns and a bandwidth of 500 MHz. The first chirp is applied at lower and the second one at higher frequencies relative to the observation position. With such a pump pulse the position of the observation frequency is not limited anymore to the edges of the nitroxide spectrum. Thus the changes in the extracted three-spin and two-spin contribution could be examined over the whole spectrum.

To prove that the duration of the double chirp is not too long and no high frequency components are lost, figure 5.22 shows experimental DEER data measured on the reference compound MSA347 with 250 MHz pump chirps. One trace was measured with a 48 ns chirp, the other one with a 96 ns chirp. The pulse positions, the external magnetic field strength and the resonator profile were identical for both experiments. The modulation depth of the two traces is identical and the real part of the phase-corrected DEER signal as well as the dipolar spectra are identical. Consequently a pulse length of 96 ns is not too long for DEER experiments on the investigated type of nitroxide triradicals.

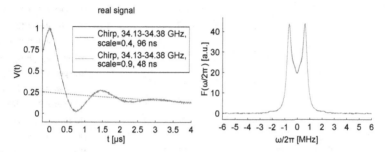

Figure 5.22: Comparison of two DEER traces with the same modulation depth measured on MSA347 with identical experimental settings but different pulse length and amplitude. Left figure: Real part of the phase-corrected DEER signals (dotted lines: fitted background). Right figure: Dipolar spectra (= Fourier transformation of the corresponding form factor).

5.1.3.1 Triradical T36d

To investigate the changes in the extracted three-spin and two-spin contribution over the whole spectrum, the pulse positions were set up such that the pulses are in the frequency range where they are most effectively coupled into the resonator. The magnetic field was chosen such that the observation frequency is positioned at the right edge of the nitroxide spectrum. With this setup, four DEER measurements with four differently scaled double chirp pump pulses (digital amplitudes of 1.0, 0.38, 0.28 and 0.127) were performed. After this the magnetic field was changed by 5 G such that the nitroxide spectrum moves to higher frequencies. With this new setup, the previously described series of DEER experiments was repeated. After this the B-field was increased by 5 G again and another series of identical DEER experiments was performed. This procedure was repeated all over and in total 17 identical series consisting of four DEER experiments each were performed. The setup of those DEER experiments is shown in figure 5.23.

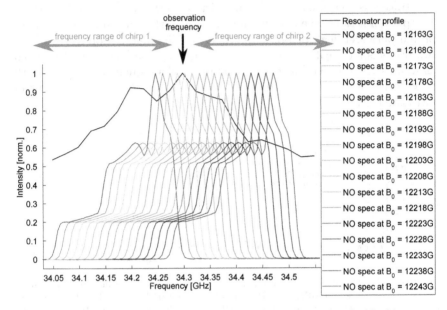

Figure 5.23: Experimental setup of the DEER experiments with a double chirp pump pulse on T36d. Shown is the position of the nitroxide absorption spectrum relative to the resonance profile of the resonator at a given external static magnetic field B_0 and how the pulse positions were set up. The double chirp consists of chirp 1 and 2.

Figure 5.24 shows the position of the observation frequency relative to the nitroxide absorption spectrum for all used magnetic field strengths. There it becomes clear that with this procedure one is able to generate comparable DEER data at observation position uniformly distributed all over the nitroxide spectrum, from the right edge to the left edge. The experimental details are summarized in table 5.4. The DEER signals for all 17 experiment series can be found in chapter 2.1 in the appendix on page 95.

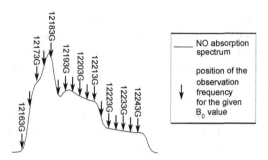

Figure 5.24: Position of the observation frequency relative to the nitroxide spectrum for specific magnetic field strengths for the DEER double-chirp experiments on T36d. The spectrum is shown in the magnetic field domain for a constant observation frequency of 34.3 GHz. Most of the other nitroxide absorption spectra in this report are shown in the frequency domain for a fixed magnetic field strength.

Table 5.4: Experimental setup of the DEER experiments with a double chirp pump pulse on T36d.

parameter	experimental setup
external magnetic field	17 different experiment series with B_0 values from 12163 G to 12243 G, increment 5 G
observation frequency	34.3 GHz
frequency range of chirp 1	34.02-34.27 GHz
frequency range of chirp 2	34.33-34.58 GHz
length of chirp 1 and 2	48 ns each

The modulation depths of the single DEER experiments are displayed in figure 5.25 **A**. The modulation depth has for all chirp amplitudes a minimum between 12173 and 12193 G. This corresponds to those series of experiments, in which the observation frequency is at the maximum of the nitroxide spectrum. At this position, the maximum is not excited by the pump chirp in those experiments. Thus, the modulation has to decrease in comparison with the other experiments, as fewer spin are pumped. Moreover, the $m_I = +1$ and $m_I = 0$ sub-spectra overlap at the observation positions for 12183 to 12193 G (compare figure 2.17 on page 33). Hence, spins whose z-axis (sub-spectra $m_I(nucleus1) = +1$) and whose x-axis are oriented parallel to the external magnetic field (sub-spectra $m_I(nucleus2) = 0$) are observed simultaneously in those cases (compare figure 2.5 on page 14). If the orientations of the spins were correlated for T36d, it might be that two spins in the same molecule are in the described orientations and are therefore observed simultaneously in 3 of 3^3 cases ($m_I(nucleus1) = +1$ and $m_I(nucleus2) = 0$ while $m_I(nucleus3) = -1, 0, +1$). If two spins are observed simultaneously, like for the described case, only one pumped spin remains. Thus, if spin 1 is observed, the inversion efficiency of spin 2 is $\frac{3}{27}$ smaller than for spin 3. To find out, how likely the discussed case is for the investigated molecules, further simulations, which sample the entire conformational space, would have to be performed.

The amplitudes of the imaginary part of the phase-corrected DEER signal are summarized in figure 5.25 **B**. The first point to mention is that the intensity of the imaginary signal is significantly smaller than in the experiments on T36d with one chirp. Moreover, the behavior of the amplitude of the imaginary signal is completely different than the one of the modulation depth. It is decreasing with increasing magnetic field strength. Hence, the imaginary signal is not directly related to the modulation depth for these double-chirp experiments.

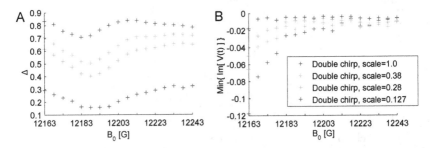

Figure 5.25: Behavior of the DEER signal of T36d for different B_0 values and double chirps as pump pulses (for the original data see chapter 2.1 in the appendix on page 95). **A)** Modulation depth. **B)** Maximal amplitude (here the minimum value of the imaginary DEER trace as the amplitude is negative) of the phase-corrected imaginary DEER signal.

The extracted two-spin and three-spin contributions are shown in figure 5.26. The time-domain signals of the two-spin and three-spin contribution (graph **A** and **B**) show a different decay rate for the different observation positions. This is related to a varying weighting of the single-frequency contributions, which is a clear hint at orientation selection [21].

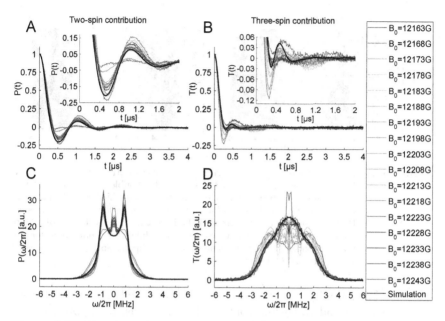

Figure 5.26: Experimental and simulated two-spin and three-spin contributions of T36d for different B_0 values and double chirps as pump pulses (for experimental details see table 5.4). Both inter-spin vectors were set to 3.64 nm, γ was set to 60° and a standard deviation of 0.25 nm was assumed. **A)** Time-domain signal of the two-spin contribution. **B)** Time-domain signal of the three-spin contribution. **C)** Dipolar spectra of the two-spin contribution. **D)** Dipolar spectra of the three-spin contribution.

The differences between the single observation positions are much bigger for T than for P. The dipolar spectra (graph **C** and **D**) confirm this observation. The differences in the frequency domain between the single two-spin contributions are small in comparison to the ones between the single three-spin contributions. The simulation of T and P fits best with the experimental data at a magnetic field strength of 12238 G.

Pseudocolor plots of the dipolar spectra of P and T versus the magnetic field strength are presented in figure 5.27. In the pseudocolor layout, the differences between the amount of variations in the two-spin and three-spin contribution become very clear. $T(\omega/2\pi)$ shows distinct differences in the frequency weighting while the characteristic frequencies in $P(\omega/2\pi)$ stay constant and only their intensities vary.

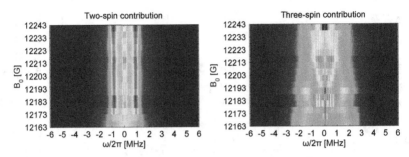

Figure 5.27: Pseudocolor plot of the experimental dipolar spectra of the two-spin and three-spin contribution of T36d versus B_0 (double chirps were used as pump pulses, for further experimental details see table 5.4). The intensity values are displayed in a color code. Color code for the print version: Black is used for the highest and the lowest intensity values. Color code for the e-book: Dark red is used for the highest and dark blue for the lowest intensity values. For the individual dipolar spectra per B_0 value see figure 5.26 **C** and **D**.

This observation might be related to assumptions made for the extraction of the two-spin and three-spin contribution. For this, the approximation was made that the inversion efficiency λ is equal for every pumped spin and independent from their orientation. The experimental results however give rise to the conclusion that this approximation is not justified for the investigated systems. Possible reasons for this were mentioned above, namely the frequency dependency in λ values and the possibility to observe two spins simultaneously, which is referred to as pump orientation selection. However, in the extraction of P and T, a constant λ value for all pumped spins is assumed. In the formula of the form factor for one particular orientation of a three-spin system, $F_3(t, \theta, \phi)$, (equation 2.48) the three-spin contribution, $T(t, \theta, \phi)$ (equation 2.53) is weighted with the product of the two inversion efficiencies $(\lambda_m \lambda_l)$ of the two pumped spins (L and M), while $P(t, \theta, \phi)$ (equation 2.52) is weighted with a combination of the single inversion efficiencies and their products $(\lambda_l - \lambda_m \lambda_l)$. If one assumes a constant λ value for all spins in the extraction of P and T, while the true λ values of the individual spins differ, the error will be bigger for those terms that are multiplied by the product of the two λ values. As the three-spin contribution is only weighted with such a product and the two-spin contribution has an additional contribution by a linear term, the error in the extraction will be bigger for the three-spin than for the two-spin contribution.

5.1.3.2 Triradical TR011

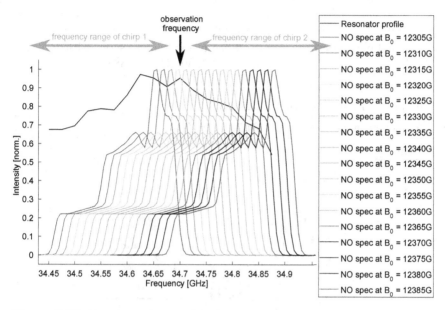

Figure 5.28: Experimental setup of the DEER experiments with a double chirp pump pulse on TR011. Shown is the position of the nitroxide absorption spectrum relative to the resonance profile of the resonator at a given external static magnetic field B_0 and the setup of the pulse positions.

On triradical TR011, DEER measurements with a double chirp pump pulse were performed at in total 17 different magnetic field strengths. The positions of the observation frequency relative to the nitroxide absorption spectrum are visualized for all used magnetic field strengths in figure 5.29. At every observation position, four DEER traces were measured with four differently scaled (1.0, 0.38, 0.28, 0.127) double chirps as pump pulses. The experimental setup is visualized in figure 5.28 and the experimental details are given in table 5.5.

Table 5.5: Experimental setup of the DEER experiments with a double chirp pump pulse on TR011.

parameter	experimental setup
external magnetic field	17 different experiment series with B_0 values from 12305 G to 12385 G
observation frequency	34.7 GHz
frequency range of chirp 1	34.42-34.67 GHz
frequency range of chirp 2	34.73-34.98 GHz
length of chirp 1 and 2	48 ns each

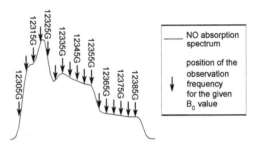

Figure 5.29: Position of the observation frequency (34.7 G) relative to the nitroxide spectrum for specific magnetic field strengths for the DEER double-chirp experiments on TR011.

The DEER signals for all 17 experiment series can be found in chapter 2.2 in the appendix on page 103. For comparability reasons, the modulation depths of the individual DEER experiments are summarized in figure 5.30 **A**. For all chirp amplitudes, a minimum is observed between 12315 and 12345 G. This is analogous to the behavior of the modulation depth in the double-chirp experiments on T36d. The decrease appears again in those series of experiments in which the observation frequency is at the maximum of the nitroxide spectrum. As the conformational flexibility of TR011 is much smaller than for T36d (see chapter 2.4.1), it is less likely that TR011 has a conformation in which the x-axis of one spin and the z-axis of another spin are parallel. Therefore, the decrease in modulation depth is probably not caused due to double observation.

The amplitudes of the imaginary part of the phase-corrected DEER signal are summarized in figure 5.30 **B**. The behavior of the amplitude of the imaginary signal can only be evaluated roughly due to experimental uncertainty (apparent kink in the data rows between 12355 and 12365 G). It takes its lowest values for the measurement at magnetic field strengths between 12325 and 12345 G. This trend agrees with the finding in the double-chirp experiments on T36d. Furthermore for most experiments, its intensity is significantly smaller than in the single chirp measurements on TR011.

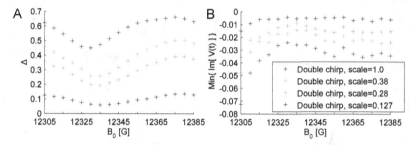

Figure 5.30: Behavior of the DEER signal of TR011 for different B_0 values and double chirps as pump pulses (for the original data see chapter 2.2 in the appendix on page 103). **A)** Modulation depth. **B)** Maximal amplitude (here the minimum value of the imaginary DEER trace as the amplitude is negative) of the phase-corrected imaginary DEER signal.

The extracted two-spin and three-spin contributions are shown in figure 5.31. The time-domain signals of the two-spin contribution (graph **A**) look pretty similar for all magnetic field strengths, whereas there are huge differences in the decay rate for the different three-spin contributions (graph **B**). The pseudocolor plots (figure 5.32) confirm these observations. The dipolar spectra (graph **C** and **D**) show that there are systematic differences in $P(\omega/2\pi)$. At low magnetic field (observation at the left edge of the nitroxide spectrum in the magnetic field domain) the spectrum is broader and flatter than the simulation, whereas at high magnetic field (observation at the right edge of the nitroxide spectrum in the magnetic field domain) the spectrum is sharper and higher than the simulation. This was also observed for T36d (compare figure 5.26). As mentioned above, pump orientation selection is less probable for TR011 than for T36d, due to less conformational flexibility. Therefore, it is likely that the changes of the experimental DEER data at different observation positions are caused by the observation of different orientations of the inter-spin vector with respect to the external magnetic field. This is referred to as orientation selection due to observer position. Observer orientation selection would lead to a deviance of the orientation weighting of the inter-spin vector from $\sin(\theta_{dd})$. The weighting would be different for every observation position.

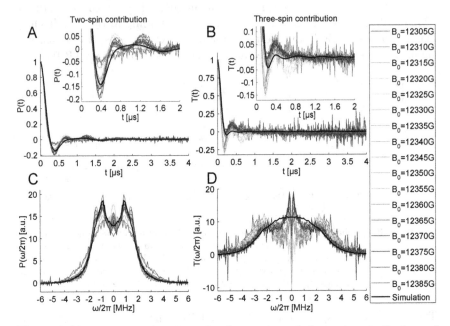

Figure 5.31: Experimental and simulated two-spin and three-spin contributions of TR011 for different B_0 values and double chirps as pump pulses (for experimental details see table 5.5). Both inter-spin vectors were set to 3.15 nm, γ was set to $72°$ and a standard deviation of 0.25 nm was assumed. **A)** Time-domain signal of the two-spin contribution. **B)** Time-domain signal of the three-spin contribution. **C)** Dipolar spectra of the two-spin contribution. **D)** Dipolar spectra of the three-spin contribution.

This would also explain the systematic changes in the two-spin spectra. Besides, the relative intensity of the two maxima in the dipolar two-spin spectra of TR011 (figure 5.31, graph **C**) changes for different observation positions. Each maximum corresponds to one of the two different inter-spin distances in TR011. This observation supports the hypotheses that the weighting of different orientations of the inter-spin vectors varies for different observation positions.

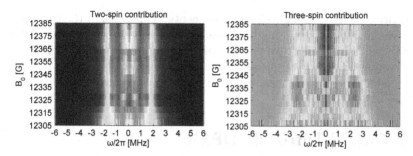

Figure 5.32: Pseudocolor plot of the experimental dipolar spectra of the two-spin and three-spin contribution of TR011 versus B_0 (double chirps were used as pump pulses, for further experimental details see table 5.5). The intensity values are displayed in a color code. Color code for the print version: Black is used for the highest and the lowest intensity values. Color code for the e-book: Dark red is used for the highest and dark blue for the lowest intensity values. For the individual dipolar spectra per B_0 value see figure 5.31 **C** and **D**.

Moreover, observer orientation selection also explains why the changes in the three-spin contribution are more significant than for the two-spin contribution. The reason for this is that the two-spin contributions of the different inter-spin vectors can be separated and each is multiplied with the corresponding weighting independently, while the three-spin contribution has to be multiplied with a combination of the different weightings. Hence, the deviation from the ideal case is roughly squared.

The simulation of T and P fits best with the experimental data at a magnetic field strength of 12380 G. This is exactly what was observed for T36d.

5.1.3.3 Conclusion

In the previous section it became clear that it is impossible to measure the three-spin contribution precisely enough for extracting the angular information in systems with three nitroxide labels. The reason for this is mainly observer orientation selection and ancillary pump orientation selection. Orthogonal spin labels might help to solve the problem of pump orientation selection. As if the spectra of the pump and observation spins do not overlap, all pumped spins can be inverted homogeneously. To avoid observer orientation selection, a spin label with much less pronounced orientation selection needs to be used.

Furthermore, it was found that the intensity of the imaginary signal is significantly reduced in the double-chirp experiments for both triradicals in comparison with the single chirp experiments. This leads to the conclusion that the imaginary signal might be caused by the pump chirp pulse and its sign depends on the relative position of the pump pulse with respect to the observation frequency. If this assumption was true, the two pump pulses in the double chirp experiment would be likely to compensate the imaginary signals of each other.

5.2 Gd(III)-Nitroxide DEER

5.2.1 DEER with a Chirp Pump Pulse

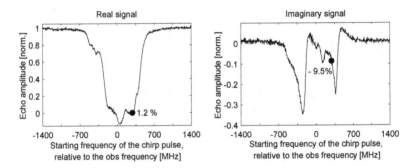

Figure 5.33: Echo reduction in Gd(III)-nitroxide DEER: Shown is the relative amplitude of the unprocessed echo in dependency of the starting frequency of a 48 ns, 250 MHz chirp pulse relative to the observation frequency. The observation frequency was set to 34.000 GHz for this experiment. The black dot indicates the starting frequency that was chosen for the DEER experiments. For a spacing of 300 MHz between the observation and the chirp starting frequency, the real part of the echo amplitude is reduced to 1.2 % of its absolute maximum value. The amplitude of the imaginary part is - 9.5 % of the maximum absolute echo amplitude.

To avoid orientation selection another model system was tried, a Gd(III)-nitroxide radical. As mentioned before (see chapter 2.4.2 and 3) Gd(III) and the nitroxides can be detected separately at Q band. Hence, the whole nitroxide spectrum can be pumped and orientation selection of the pump pulse can be avoided completely. Moreover, Gd(III) is known to have a much less pronounced orientation selection due to the zero-field splitting [25]. Thus, this systems seems to be promising to exclude orientation selection. However, the echo reduction effect for Gd(III)(see chapter 2.3.3.2) may be a problem. Therefore the echo amplitude was measured for different chirp positions. Figure 5.33 shows the real and imaginary part of the amplitude of the unprocessed (non phase-corrected) DEER echo in dependency of the chirp position. If the spacing between the chirp pulse and the observation frequency is too small, hardly any signal can be detected. However, the spacing is limited by the spectral range of the Gd(III) absorption spectrum and the bandwidth of the resonator. Therefore, a compromise had to be found between all three parameters. As the position of the chirp pump pulse is set by the nitroxide absorption spectrum, the observation frequency has to be moved away from the central transition of the Gd(III) spectra. It was placed 300 MHz away from the pump pulse, at 34.00 GHz (compare figure 5.34). With this spacing an echo reduction of 98.8 % in the real signal and an imaginary signal of -9.5 % of the absolute maximum echo amplitude could be achieved. It has to be mentioned that these values are not corrected for the phase. Hence, it is likely that most of the imaginary signal is transfered to the real signal by phase correction.

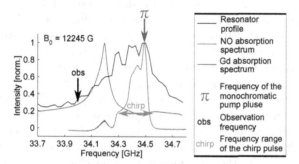

Figure 5.34: Experimental setup of the DEER experiments on the Gd(III)-nitroxide radical: Shown is the position of the Gd(III) and the nitroxide absorption spectrum relative to the resonance profile of the resonator at a given external static magnetic field B_0 and how the pulse positions were set up.

Table 5.6: Experimental setup of the DEER experiments on the Gd(III)-nitroxide radical.

parameter	experimental setup
external magnetic field	12245 G
observation frequency	34.00 GHz
frequency range of the chirp pump pulse	34.30-34.55 GHz
length of the chirp pump pulse	48 ns
frequency of the monochromatic pump pulse	34.49 GHz
length of the monochromatic pump pulse	17 ns

In total three DEER traces were measured with this pulse setting (for details see table 5.6). The amplitude of the used chirp pump pulse was different for every experiment, the used digital scales were 1.0, 0.11 and 0.01. The analog amplitude of the last pump pulse was additionally reduced by 10 dB. Furthermore one DEER trace with a π pump pulse was measured.

The experimental DEER data is shown in figure 5.35. With the most intense chirp pulse a modulation depth of 89 % could be achieved, which is identical to the saturation value (compare figure 4.2). The real part of the phase-corrected DEER signals is presented in graph **A**. The traces that were measured with the chirp pulses and the one of the monochromatic pump pulse have all a well-defined starting point. Different modulation depths could be achieved. The amplitude of the phase-corrected imaginary signal (graph **B**) scales with the pulse amplitude of the chirps, respectively with the modulation depth. There are big differences in the form factors (compare graph **C** and **D**). All four show a kink at around 0.1 μs. However, this kink gets more pronounced with increasing modulation depth. It is thus caused by the three-spin contribution. Moreover, the decay of the form factors becomes faster with increasing modulation depth. This means that the contribution of high frequencies gets higher. This finding is also confirmed by the dipolar spectra (graph **E**).

The experimental and simulated two-spin and three-spin contribution are presented in figure 5.36. Here the simulation and the experimental data fit very well for both the two-spin and the three-spin contribution. Appropriate experimental conditions, that allow the extraction of the three-spin contribution with the assumptions made, have thus been found.

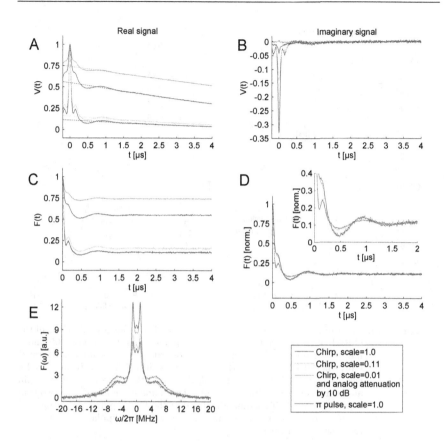

Figure 5.35: Experimental DEER data for the Gd(III)-nitroxide radical (for the experimental setup see table 5.6). **A)** Real part of the phase-corrected DEER signals (dotted lines: fitted background). **B)** Imaginary part of the phase-corrected DEER signals. **C)** Form factors (= phase- and background-corrected experimental DEER signals, dotted lines: fit from which the dipolar spectra is calculated). **D)** Normalized form factors (= phase- and background-corrected experimental DEER signals that are scaled such that they have the same modulation depth). **E)** Dipolar spectra (= Fourier transformation of the corresponding form factor).

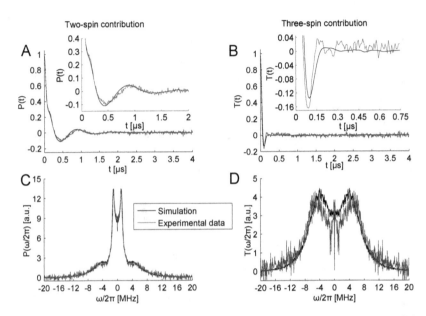

Figure 5.36: Experimental and simulated two-spin and three-spin contributions of the Gd(III)-nitroxide radical. The inter-spin vectors were set to 3.5 and 2 nm, γ was set to 20° and a standard deviation of 0.25 nm was assumed. Only the DEER traces of the experiments in which a chirp pump pulse was used were include in the calculation of the experimental two-spin and three-spin contribution. **A)** Time-domain signal of the two-spin contribution. **B)** Time-domain signal of the three-spin contribution. **C)** Dipolar spectra of the two-spin contribution. **D)** Dipolar spectra of the three-spin contribution.

5.2.2 Conclusion

It was found that it is, in principle, possible to avoid orientation selection by using a Gd(III) label as observer spin in combination with nitroxide labels as pumped spins. However, this could only be realized at the expense of intensity losses due to echo reduction and faster relaxation times, as the observer position had to be moved away from the central transition. Such conditions are clearly not the optimum for extracting the three-spin contribution, as the signal-to-noise ratio is poor. However, it is nowadays not possible due to technical limitations to generate chirps with enough power and resonators with enough bandwidth to do the measurement at higher field strength, where the spacing between the absorption spectra of Gd(III) and nitroxide is larger. Another option to increase the spacing between Gd(III) and nitroxide absorption spectra is to develop other Gd(III) ligands that reduce the zero-field splitting and thus lead to a sharper Gd(III) spectrum.

6 Summary and Outlook

The aim of this thesis was to find experimental conditions under which the three-spin contribution can be extracted precisely from the DEER signal for a system with three spins.

As a starting point, simulations of the two-spin and three-spin contribution were performed for nitroxide triradicals. These investigations showed that the changes in the three-spin contribution, which are directly related to variations in the angle, have about the same magnitude as those changes in the three- and two-spin contribution that are caused by variations in the third side length. From this result the conclusion was drawn that DEER data with very high resolution is needed. Hence the measurements were performed on Q band.

In the theoretical considerations, it became clear that several DEER traces with varying modulation depths are required to separate the two-spin and three-spin contribution precisely. Moreover these considerations assumed a homogeneous excitation of the pump spins over the whole spectrum. This was achieved by using chirp pulses with a huge bandwidth, produced by a fast arbitrary waveform generator (AWG), as pump pulses. Besides this a homemade resonator with a sufficient bandwidth was used in a highly overcoupled state. With this experimental setup high modulation depths ($> 80\%$) could be achieved.

First experiments with one chirp pulse or a double chirp pulse, consisting of two separate chirps, were performed on triply labeled nitroxide radicals. The result of the investigations on the nitroxide triradicals was that, due to orientation selection, it is not possible for those radicals to extract the two-spin and the three-spin contribution from the experimental DEER data with the assumptions made. The differences in the frequency weighting due to orientation selection were much more pronounced for T than for P. In addition non-zero phase-corrected imaginary signals could be observed. Further experiments performed on a nitroxide diradical showed that the imaginary signals might be caused by the pump pulse. Besides it appears that the real part of the phase-corrected DEER signals is not affected by the non-zero phase-corrected imaginary signal.

To solve the problem of orientation selection a Gd(III)-nitroxide radical was introduced as a model system. The absorption spectra of Gd(III) and the nitroxide labels do partially overlap at Q band. As the whole nitroxide spectrum was pumped, this caused a reduction of the echo amplitude of the observed Gd(III) species. To cope with this the observation frequency was moved away from the central transition of the Gd(III). With this compound the experimental and simulated two-spin and three-spin contribution matched very well. The appropriate experimental conditions and way of labeling have thus been found to perform the extraction of the three-spin contribution with the assumptions made.

In a next step, a new analysis procedure needs to be developed to extract the angular information out of the extracted three-spin contribution. Further one could think of expanding the methods on multiple-spin systems. This is important, as in some cases, e.g. for homooligomeric proteins, it is not possible to produce samples containing only two pump spin labels and one observer spin label. [9] For systems with only one observer spin, it should be possible to extract the higher-order terms with the procedure introduced in this thesis. Systems with more than one observer spin, however, are more challenging and more investigations are needed.

Additional Spectra

1 Simulation of the Angle Dependency of the Three-Spin and the Two-Spin Contribution

1.1 Simulation with a Fix Observer Spin

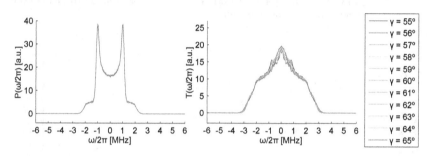

Figure 1: Simulated dipolar spectra of the two-spin and three-spin contribution of T36d with fixed observer spin: Both inter-spin vectors were set to 3.64 nm with a standard deviation of 0.1 nm.

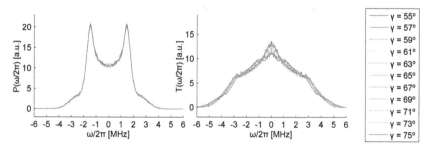

Figure 2: Simulated dipolar spectra of the two-spin and three-spin contribution of TR011 with fixed observer spin: Both inter-spin vectors were set to 3.15 nm with a standard deviation of 0.15 nm.

1.2 Simulation with a Variable Observer Spin

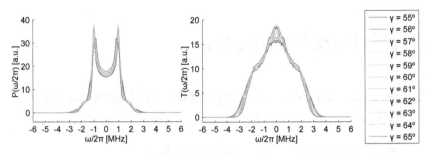

Figure 3: Simulated dipolar spectra of the two-spin and three-spin contribution of T36d with variable observer spin: Both inter-spin vectors were set to 3.64 nm with a standard deviation of 0.1 nm.

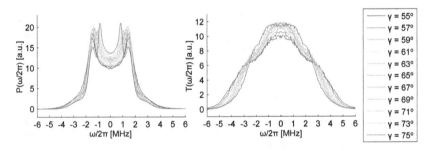

Figure 4: Simulated dipolar spectra of the two-spin and three-spin contribution of TR011 with variable observer spin: Both inter-spin vectors were set to 3.15 nm with a standard deviation of 0.15 nm.

2 DEER with a Double Chirp as Pump Pulse

2.1 Triradical T36d

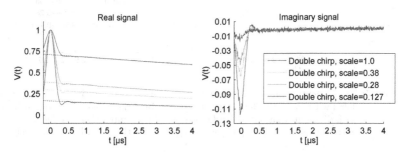

Figure 5: Experimental, phase-corrected DEER signals (real and imaginary part) of T36d with $B_0 = 12163$ G (for details see table 5.4 on page 75). Double chirps with different digital amplitudes were used as pump pulses (see legend). The dotted lines in the real part of the signal show the fitted background signal.

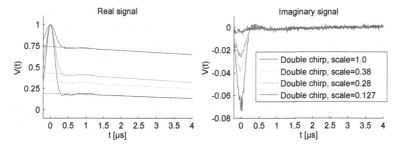

Figure 6: Experimental, phase-corrected DEER signals (real and imaginary part) of T36d with $B_0 = 12168$ G (for details see table 5.4 on page 75). Double chirps with different digital amplitudes were used as pump pulses (see legend). The dotted lines in the real part of the signal show the fitted background signal.

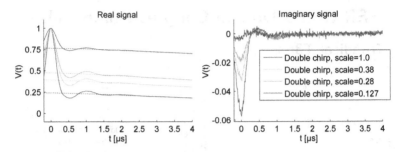

Figure 7: Experimental, phase-corrected DEER signals (real and imaginary part) of
T36d with $B_0 = 12173$ G (for details see table 5.4 on page 75). Double chirps
with different digital amplitudes were used as pump pulses (see legend). The
dotted lines in the real part of the signal show the fitted background signal.

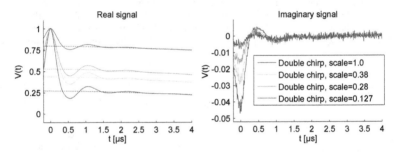

Figure 8: Experimental, phase-corrected DEER signals (real and imaginary part) of
T36d with $B_0 = 12178$ G (for details see table 5.4 on page 75). Double chirps
with different digital amplitudes were used as pump pulses (see legend). The
dotted lines in the real part of the signal show the fitted background signal.

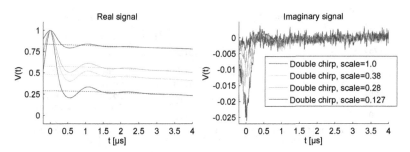

Figure 9: Experimental, phase-corrected DEER signals (real and imaginary part) of T36d with $B_0 = 12183$ G (for details see table 5.4 on page 75). Double chirps with different digital amplitudes were used as pump pulses (see legend). The dotted lines in the real part of the signal show the fitted background signal.

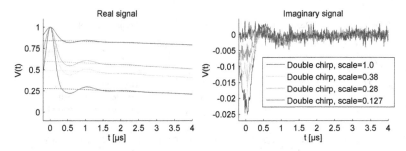

Figure 10: Experimental, phase-corrected DEER signals (real and imaginary part) of T36d with $B_0 = 12188$ G (for details see table 5.4 on page 75). Double chirps with different digital amplitudes were used as pump pulses (see legend). The dotted lines in the real part of the signal show the fitted background signal.

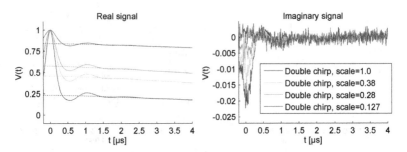

Figure 11: Experimental, phase-corrected DEER signals (real and imaginary part) of T36d with $B_0 = 12193$ G (for details see table 5.4 on page 75). Double chirps with different digital amplitudes were used as pump pulses (see legend). The dotted lines in the real part of the signal show the fitted background signal.

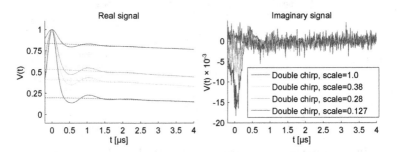

Figure 12: Experimental, phase-corrected DEER signals (real and imaginary part) of T36d with $B_0 = 12198$ G (for details see table 5.4 on page 75). Double chirps with different digital amplitudes were used as pump pulses (see legend). The dotted lines in the real part of the signal show the fitted background signal.

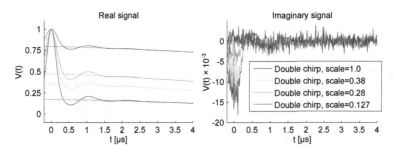

Figure 13: Experimental, phase-corrected DEER signals (real and imaginary part) of T36d with $B_0 = 12203$ G (for details see table 5.4 on page 75). Double chirps with different digital amplitudes were used as pump pulses (see legend). The dotted lines in the real part of the signal show the fitted background signal.

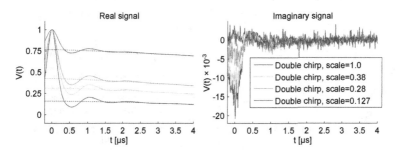

Figure 14: Experimental, phase-corrected DEER signals (real and imaginary part) of T36d with $B_0 = 12208$ G (for details see table 5.4 on page 75). Double chirps with different digital amplitudes were used as pump pulses (see legend). The dotted lines in the real part of the signal show the fitted background signal.

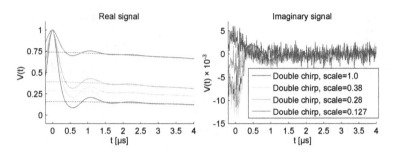

Figure 15: Experimental, phase-corrected DEER signals (real and imaginary part) of T36d with $B_0 = 12213$ G (for details see table 5.4 on page 75). Double chirps with different digital amplitudes were used as pump pulses (see legend). The dotted lines in the real part of the signal show the fitted background signal.

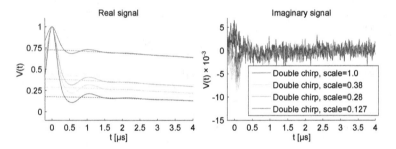

Figure 16: Experimental, phase-corrected DEER signals (real and imaginary part) of T36d with $B_0 = 12218$ G (for details see table 5.4 on page 75). Double chirps with different digital amplitudes were used as pump pulses (see legend). The dotted lines in the real part of the signal show the fitted background signal.

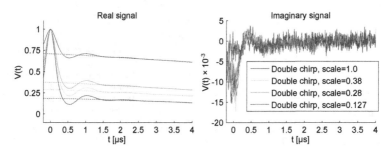

Figure 17: Experimental, phase-corrected DEER signals (real and imaginary part) of
T36d with $B_0 = 12223$ G (for details see table 5.4 on page 75). Double
chirps with different digital amplitudes were used as pump pulses (see legend). The dotted lines in the real part of the signal show the fitted background signal.

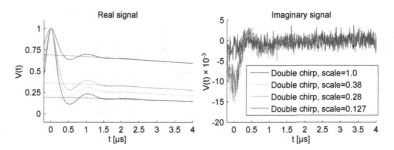

Figure 18: Experimental, phase-corrected DEER signals (real and imaginary part) of
T36d with $B_0 = 12228$ G (for details see table 5.4 on page 75). Double
chirps with different digital amplitudes were used as pump pulses (see legend). The dotted lines in the real part of the signal show the fitted background signal.

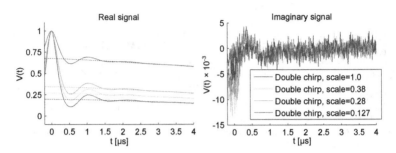

Figure 19: Experimental, phase-corrected DEER signals (real and imaginary part) of T36d with $B_0 = 12233$ G (for details see table 5.4 on page 75). Double chirps with different digital amplitudes were used as pump pulses (see legend). The dotted lines in the real part of the signal show the fitted background signal.

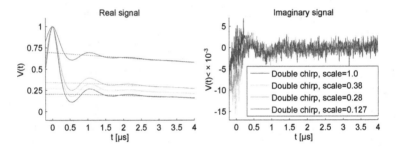

Figure 20: Experimental, phase-corrected DEER signals (real and imaginary part) of T36d with $B_0 = 12238$ G (for details see table 5.4 on page 75). Double chirps with different digital amplitudes were used as pump pulses (see legend). The dotted lines in the real part of the signal show the fitted background signal.

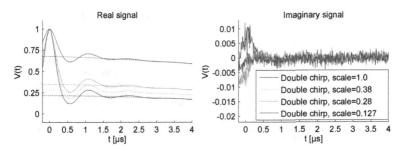

Figure 21: Experimental, phase-corrected DEER signals (real and imaginary part) of T36d with $B_0 = 12243$ G (for details see table 5.4 on page 75). Double chirps with different digital amplitudes were used as pump pulses (see legend). The dotted lines in the real part of the signal show the fitted background signal.

2.2 Triradical TR011

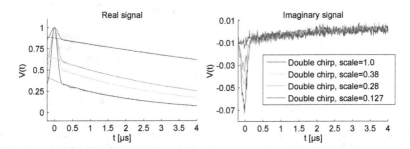

Figure 22: Experimental, phase-corrected DEER signals (real and imaginary part) of TR011 with $B_0 = 12305$ G (for details see table 5.5 on page 80). Double chirps with different digital amplitudes were used as pump pulses (see legend). The dotted lines in the real part of the signal show the fitted background signal.

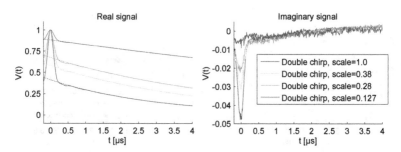

Figure 23: Experimental, phase-corrected DEER signals (real and imaginary part) of TR011 with $B_0 = 12310$ G (for details see table 5.5 on page 80). Double chirps with different digital amplitudes were used as pump pulses (see legend). The dotted lines in the real part of the signal show the fitted background signal.

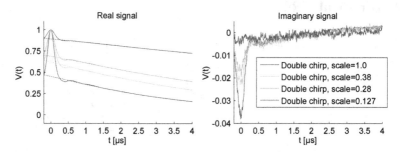

Figure 24: Experimental, phase-corrected DEER signals (real and imaginary part) of TR011 with $B_0 = 12315$ G (for details see table 5.5 on page 80). Double chirps with different digital amplitudes were used as pump pulses (see legend). The dotted lines in the real part of the signal show the fitted background signal.

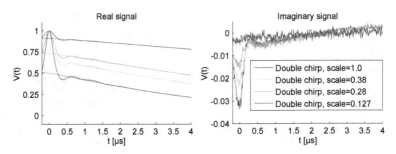

Figure 25: Experimental, phase-corrected DEER signals (real and imaginary part) of TR011 with $B_0 = 12320$ G (for details see table 5.5 on page 80). Double chirps with different digital amplitudes were used as pump pulses (see legend). The dotted lines in the real part of the signal show the fitted background signal.

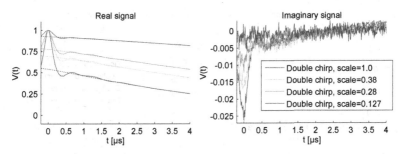

Figure 26: Experimental, phase-corrected DEER signals (real and imaginary part) of TR011 with $B_0 = 12325$ G (for details see table 5.5 on page 80). Double chirps with different digital amplitudes were used as pump pulses (see legend). The dotted lines in the real part of the signal show the fitted background signal.

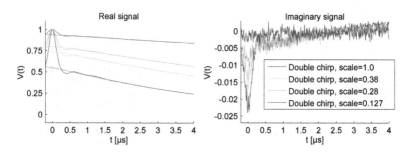

Figure 27: Experimental, phase-corrected DEER signals (real and imaginary part) of
TR011 with $B_0 = 12330$ G (for details see table 5.5 on page 80). Double
chirps with different digital amplitudes were used as pump pulses (see legend). The dotted lines in the real part of the signal show the fitted background signal.

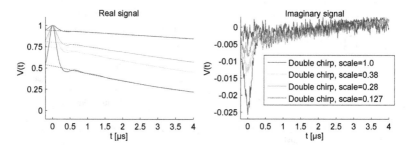

Figure 28: Experimental, phase-corrected DEER signals (real and imaginary part) of
TR011 with $B_0 = 12335$ G (for details see table 5.5 on page 80). Double
chirps with different digital amplitudes were used as pump pulses (see legend). The dotted lines in the real part of the signal show the fitted background signal.

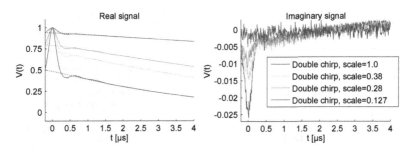

Figure 29: Experimental, phase-corrected DEER signals (real and imaginary part) of TR011 with $B_0 = 12340$ G (for details see table 5.5 on page 80). Double chirps with different digital amplitudes were used as pump pulses (see legend). The dotted lines in the real part of the signal show the fitted background signal.

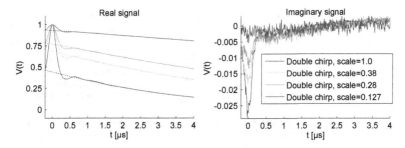

Figure 30: Experimental, phase-corrected DEER signals (real and imaginary part) of TR011 with $B_0 = 12345$ G (for details see table 5.5 on page 80). Double chirps with different digital amplitudes were used as pump pulses (see legend). The dotted lines in the real part of the signal show the fitted background signal.

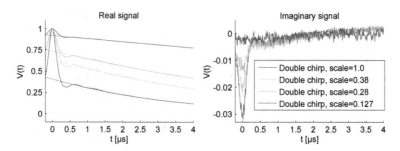

Figure 31: Experimental, phase-corrected DEER signals (real and imaginary part) of
TR011 with $B_0 = 12350$ G (for details see table 5.5 on page 80). Double
chirps with different digital amplitudes were used as pump pulses (see legend). The dotted lines in the real part of the signal show the fitted background signal.

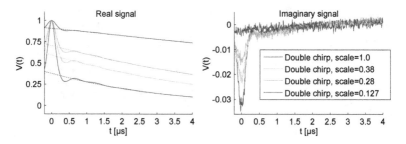

Figure 32: Experimental, phase-corrected DEER signals (real and imaginary part) of
TR011 with $B_0 = 12355$ G (for details see table 5.5 on page 80). Double
chirps with different digital amplitudes were used as pump pulses (see legend). The dotted lines in the real part of the signal show the fitted background signal.

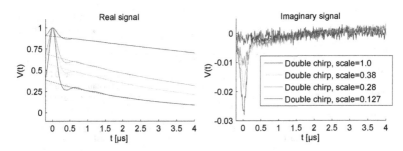

Figure 33: Experimental, phase-corrected DEER signals (real and imaginary part) of TR011 with $B_0 = 12360$ G (for details see table 5.5 on page 80). Double chirps with different digital amplitudes were used as pump pulses (see legend). The dotted lines in the real part of the signal show the fitted background signal.

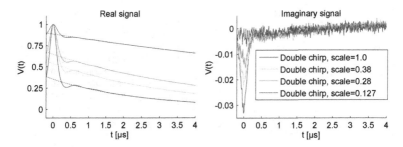

Figure 34: Experimental, phase-corrected DEER signals (real and imaginary part) of TR011 with $B_0 = 12365$ G (for details see table 5.5 on page 80). Double chirps with different digital amplitudes were used as pump pulses (see legend). The dotted lines in the real part of the signal show the fitted background signal.

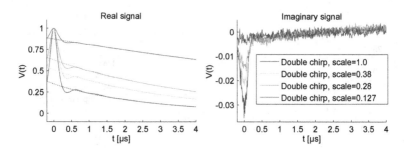

Figure 35: Experimental, phase-corrected DEER signals (real and imaginary part) of TR011 with $B_0 = 12370$ G (for details see table 5.5 on page 80). Double chirps with different digital amplitudes were used as pump pulses (see legend). The dotted lines in the real part of the signal show the fitted background signal.

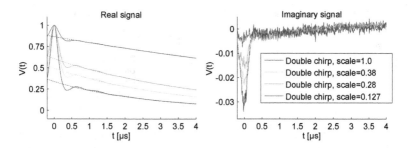

Figure 36: Experimental, phase-corrected DEER signals (real and imaginary part) of TR011 with $B_0 = 12375$ G (for details see table 5.5 on page 80). Double chirps with different digital amplitudes were used as pump pulses (see legend). The dotted lines in the real part of the signal show the fitted background signal.

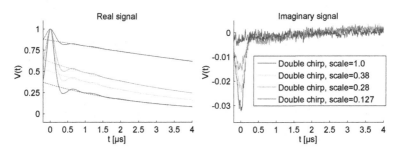

Figure 37: Experimental, phase-corrected DEER signals (real and imaginary part) of TR011 with $B_0 = 12380$ G (for details see table 5.5 on page 80). Double chirps with different digital amplitudes were used as pump pulses (see legend). The dotted lines in the real part of the signal show the fitted background signal.

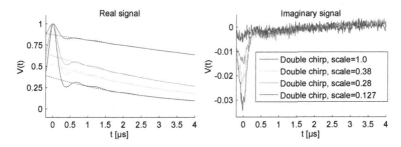

Figure 38: Experimental, phase-corrected DEER signals (real and imaginary part) of TR011 with $B_0 = 12385$ G (for details see table 5.5 on page 80). Double chirps with different digital amplitudes were used as pump pulses (see legend). The dotted lines in the real part of the signal show the fitted background signal.

Bibliography

[1] Jeschke, G., Sajid, M., Schulte, M. & Godt, A. Three-spin correlations in double electron-electron resonance. *Phys. Chem. Chem. Phys.* **11**, 6580–6591 (2009).

[2] Weil, J. A. & Bolton, J. R. *Electron Paramagnetic Resonance, Second Edition* (John Wiley & Sons, Inc., 2007).

[3] Schweiger, A. & Jeschke, G. *Principles of pulse electron paramagnetic resonance* (Oxford University Press, 2001).

[4] Eaton, G. R., Eaton, S. S., Barr, D. P. & Weber, R. T. *Quantitative EPR* (Springer, 2010).

[5] Buxbaum, E. *Fundamentals of Protein Structure and Function* (Springer, 2007).

[6] Lüders, P. *Long Range Distance Information on Orthogonal Spin Pairs by EPR.* Ph.D. thesis, ETH Zurich (2011).

[7] Swanson, M. A. *et al.* DEER Distance Measurement Between a Spin Label and a Native FAD Semiquinone in Electron Transfer Flavoprotein. *J Am Chem Soc.* **131**, 15978–15979 (2009).

[8] Klose, D. *et al.* Simulation vs. Reality: A Comparison of In Silico Distance Predictions with DEER and FRET Measurements. *PLoS ONE* **7**, e39492–e39492 (2012).

[9] von Hagens, T. *Quantification of Modulation Depth and Ghost Distance Removal in Diploar EPR on Multiple-Spin Systems.* Ph.D. thesis, ETH Zurich (2014).

[10] von Hagens, T., Polyhach, Y., Sajid, M., Godt, A. & Jeschke, G. Suppression of ghost distances in multiple-spin double electron-electron resonance. *Phys. Chem. Chem. Phys.* **15**, 5854–5866 (2013).

[11] Doll, A., Pribitzer, S., Tschaggelar, R. & Jeschke, G. Adiabatic and fast passage ultra-wideband inversion in pulsed EPR. *Journal of Magnetic Resonance* **230**, 27–39 (2013).

[12] Jeschke, G. *Kurze Einführung in die elektronenparamagnetische Resonanzspektroskopie, lecture script,* (University of Konstanz, 2008).

[13] Keeler, J. *Understanding NMR Spectroscopy, Second Edition* (John Wiley & Sons, Inc., 2010).

[14] Baum, J., Tycko, R. & Pines, A. Broadband and adiabatic inversion of a two-level system by phase-modulated pulses. *Physical Review A* **32**, 3435–3447 (1985).

[15] Garwood, M. & DelaBarre, L. The Return of the Frequency Sweep: Designing Adiabatic Pulses for Contemporary NMR. *Journal of Magnetic Resonance* **153**, 155–177 (2001).

[16] Tannús, A. & Garwood, M. Adiabatic Pulses. *NMR in Biomedicine* **10**, 423–434 (1997).

[17] Kielmann, U. *Structural Studies of Weakly Ordered Materials by Spin Probe EPR Techniques.* Ph.D. thesis, ETH ZURICH (2012).

[18] Jeschke, G. DEER Distance Measurements on Proteins. *Annu. Rev. Phys. Chem.* **63**, 419–446 (2012).

[19] Jeschke, G. *DEER = PELDOR, Presentation* (6th EF-EPR Summer School, Rehovot, 2013).

[20] Yulikov, M., Lüders, P., Warsi, M. F., Chechik, V. & Jeschke, G. Distance measurements in Au nanoparticles functionalized with nitroxide radicals and Gd^{3+}-DTPA chelate complexes. *Phys. Chem. Chem. Phys.* **14**, 10732–10746 (2012).

[21] Polyhach, Y., Godt, A., Bauer, C. & Jeschke, G. Spin pair geometry revealed by high-field DEER in the presence of conformational distributions. *Journal of Magnetic Resonance* **185**, 118–129 (2007).

[22] Banham, J. *et al.* Distance measurements in the borderline region of applicability of CW EPR and DEER: A model study on a homologous series of spin-labelled peptides. *Journal of Magnetic Resonance* **191**, 202–218 (2008).

[23] Jeschke, G. Distance Measurements in the Nanometer Range by Pulse EPR. *ChemPhysChem* **3**, 927–932 (2002).

[24] Yulikov, M. Spectroscopically orthogonal spin labels and distance measurements in biomolecules. *Electron Paramag. Reson.* **24**, 1–31 (2015).

[25] Lüders, P., Jeschke, G. & Yulikov, M. Double Electron-Electron Resonance Measured Between Gd^{3+} Ions and Nitroxide Radicals. *J. Phys. Chem. Lett.* **2**, 604–609 (2011).

[26] Garbuio, L. *et al.* Orthogonal Spin Labeling and Gd(III)-Nitroxide Distance Measurements on Bacteriophage T4-Lysozyme. *J. Phys. Chem. B* **117**, 3145–3153 (2013).

[27] Lüders, P., Jäger, H., Hemminga, M. A., Jeschke, G. & Yulikov, M. Distance Measurements on Orthogonally Spin-Labeled Membrane Spanning WALP23 Polypeptides. *J. Phys. Chem. B* **117**, 2061–2068 (2013).

[28] Ernst, M., Meier, B. H. & Jeschke, G. *Lecture script for PC IV: Magnetic Resonance* (ETH Zurich, 2012).

[29] Rinard, G. A., Quine, R. W., Eaton, S. S., Eaton, G. R. & Froncisz, W. Relative Benefits of Overcoupled Resonators vs Inherently Low-Q Resonators for Pulsed Magnetic Resonance. *Journal of Magnetic Resonance, Series A* **108**, 71–81 (1994).

[30] Jeschke, G. *Lecture script for PC VI: Signal processing in spectroscopy* (ETH Zurich, 2013).

[31] Mims, W. B. & Peisach, J. *Biological Magnetic Resonance*, vol. 3, chap. Electron Spin Echo Spectroscopy and the Study of Metalloproteins, 213–263 (Plenum Press, 1981).

[32] Garbuio, L. *Extending Gd(III)-nitroxide DEER methodology from model system to biological applications.* Ph.D. thesis, ETH Zurich (2014).

[33] Polyhach, Y. *et al.* High sensitivity and versatility of the DEER experiment on nitroxide radical pairs at Q-band frequencies. *Phys. Chem. Chem. Phys.* **14**, 10762–10773 (2012).

[34] Doll, A. & Jeschke, G. Fourier-transform electron spin resonance with bandwidth-compensated chirp pulses. *Journal of Magnetic Resonance* **246**, 18–26 (2014).

[35] Jeschke, G. *et al.* DeerAnalysis2006 - a Comprehensive Software Package for Analyzing Pulsed ELDOR Data. *Appl. Magn. Reson.* **30**, 473–498 (2006).

[36] Jeschke, G. *DeerAnalysis2013.2 User Manual.* ETH Zurich (2013).

[37] Marko, A., Denysenkov, V. & Prisner, T. F. Out-of-phase PELDOR. *Molecular Physics* **111**, 2834–2844 (2013).

[38] Bowman, M. K. & Maryasov, A. G. Dynamic phase shifts in nanoscale distance measurements by double electron electron resonance (DEER). *Journal of Magnetic Resonance* **185**, 270–282 (2007).

Printed in the United States
by Bookmasters

Printed in the United States
By Bookmasters